# Last Chance Tourism

Concerns over vanishing destinations such as the Great Barrier Reef, Antarctica, and the ice cap on Mt. Kilimanjaro have prompted some travel operators and tour agencies to recommend these destinations to consumers before they disappear. This travel trend has been reported as: "disappearing tourism," "doom tourism," and most commonly "last chance tourism," where tourists explicitly seek vanishing landscapes or seascapes, and/or disappearing natural and/or social heritage. However, despite this increasing form of travel there has been little examination in the academic literature of the last chance tourism phenomenon.

This is the first book to empirically examine and evaluate this contemporary tourism development, providing a new angle on the effects of global change and pressures of visitation on tourism destinations. It aims to develop the conceptual definition of last chance tourism, examine the ethics surrounding this type of travel, and provide case studies highlighting this form of tourism in different regions, and in different contexts. In particular it critically reviews the advantages of publicizing vulnerable destinations to raise awareness and promote conservation efforts. Conversely, the book draws attention to the issue of attracting more tourists seeking to undergo such experiences before they are gone forever, accelerating the negative impacts. It further examines current trends, discusses escalating challenges, provides management strategies, and highlights future research opportunities.

*Last Chance Tourism* is a timely and multi-disciplinary volume featuring contributions from leading scholars in the fields of leisure, tourism, anthropology, geography, and sociology. It draws on a range of international case studies and will be of interest to students, researchers, and academics interested in Tourism, Environmental Studies, and Development Studies.

**Raynald Harvey Lemelin** is the LU-SSHRC Research Chair in Parks and Protected Areas at Lakehead University, Ontario, Canada.

**Jackie Dawson** is the Canada Research Chair in Environment, Society and Policy at the University of Ottawa, Canada.

**Emma J. Stewart** is a Senior Lecturer in tourism and parks at Lincoln University, New Zealand and is a Research Associate at the Arctic Institute of North America (AINA).

# List of illustrations

## Figures

## Tables

# Contents

**As each of the editors have personal and professional ties to Aotearoa/New Zealand we would like to dedicate this book to the brave souls lost in the Christchurch earthquake – February 2011.**

# Last Chance Tourism

## Adapting tourism opportunities in a changing world

**Edited by**
**Raynald Harvey Lemelin,**
**Jackie Dawson and**
**Emma J. Stewart**

Routledge
Taylor & Francis Group
LONDON AND NEW YORK

First published 2012
by Routledge
2 Park Square, Milton Park, Abingdon, Oxon OX14 4RN

Simultaneously published in the USA and Canada
by Routledge
711 Third Avenue, New York, NY 10017

*Routledge is an imprint of the Taylor & Francis Group, an informa business*



*British Library Cataloguing in Publication Data*
A catalogue record for this book is available from the British Library

*Library of Congress Cataloging in Publication Data*
Last chance tourism / Harvey Lemelin, Jackie Dawson
and Emma J. Stewart, [editors].
p. cm.
Includes bibliographical references and index
1. Tourism—Social aspects. I. Lemelin, Harvey. II. Dawson, Jackie
III. Stewart, Emma J.
G155.A1L3755 2011
338.4'791—dc23
2011021213

ISBN: 978–0–415–61823–6 (hbk)
ISBN: 978–0–203–82893–9 (ebk)

Typeset in Times New Roman
by Swales & Willis Ltd, Exeter, Devon

# Contributors

**Bas Amelung** is a researcher at the Environmental Systems Analysis Group (ESA) at Wageningen University and Research Centre (WUR), the Netherlands. His research interests include sustainable tourism, tourism and climate change mitigation, and adaptation.

**Gary Baikie** is the Visitor Experience Coordinator for the Torngat Mountains National Park. Through his work at the Torngasoak Cultural Centre in Nain, Gary's supervised the Taipsumani Archaeological Project in Hebron, NL and help to repatriated the 113 skeletal remains back to Sallikuluk in 1996.

**Alexandra Coghlan** is a Research Fellow at Griffith University's International Centre for Ecotourism Research. Her research priorities are linking conservation with tourism, investigating a range of partnerships between NGOs, natural resource managers, communities, and tourism business. Her primary research interests are tourism partnerships, volunteer tourism, environmental education, and nature-based tourism more broadly.

**Jackie Dawson** is the Canada Research Chair in Environment, Society and Policy at the University of Ottawa where she is affiliated with the Department of Geography and the Institute for Science Society and Policy. Her research focus is on the human dimensions of global change including transforming Arctic economies.

**Paul F.J. Eagles** is a Professor in the Department of Recreation and Leisure Studies, University of Waterloo, Canada. Paul Eagles is an environmental planner with a broad interest in the application of ecology and planning theory to environmental conservation. Professionally, as a biologist and a planner, he has a long-standing interest in the planning of national parks and other forms of protected areas.

**Eke Eijgelaar** is a Researcher at the Centre for Sustainable Tourism and Transport, NHTV Breda University of Applied Sciences, the Netherlands. His research interests include tourism and climate change, tourism in protected areas, World Heritage tourism, and sustainable tourism.

**Elspeth Frew** is a Senior Lecturer in Tourism Management and a research associate of the Tourism and Hospitality Research Unit at La Trobe University, Melbourne, Australia. Her research focuses on two distinct areas – first, cultural tourism with a particular focus on festival and events management and second, the adaptation of tourist behavior due to climate change.

**Glen T. Hvenegaard** is a Professor of Environmental Science. His research focuses on protected areas, ecotourism, bird conservation, and environmental education. He is a member of the World Commission on Protected Areas and LEAD International (Leadership for Environment and Development).

**Rhonda L. Koster** is an Associate Professor in the School of Outdoor Recreation, Parks and Tourism at Lakehead University, Ontario. Her research focuses on the contribution of tourism toward rural sustainability, with expertise in the areas of building capacity for tourism development with First Nations, and, developing frameworks for evaluating tourism as a community economic development endeavor.

**Machiel Lamers** is a Researcher at the Environmental Policy Group (ENP) at Wageningen University and Research Centre (WUR), the Netherlands. His research interests include polar tourism, sustainable tourism development, tourism and nature conservation, and tourism scenarios.

**Raynald Harvey Lemelin** is the LU-SSHRC Research Chair in Parks and Protected Areas at Lakehead University, Ontario, Canada. His research interests include northern parks and the co-management of these protected areas, aboriginal tourism, and wildlife tourism. Dr. Lemelin is currently participating on a number of collaborative research projects with First Nations and Inuit communities located throughout Northern Canada.

**Christopher J. Lemieux** is a Postdoctoral Fellow at the Wilfrid Laurier University (Ontario, Canada), an Affiliate Faculty member at Colorado State University (Fort Collins, U.S.A.), and is Director of the Canadian Council on Ecological Areas. Dr. Lemieux's research focuses primarily on institutional adaptation to climate change in the protected areas sector.

**Margaret Johnston** is a Professor in Outdoor Recreation, Parks and Tourism at Lakehead University in Thunder Bay, Canada. Her research has examined Arctic tourism patterns, development and regulation. She is currently working with a team on exploring the relationships between tourism change and climate change in the Arctic.

**David Newsome** is an Associate Professor in the School of Environmental Science at Murdoch University. His teaching and research interests span many areas of natural area tourism including wildlife tourism, the biophysical impacts of recreation in protected areas, evaluation of the quality of ecotourism operations, sustainable trail management, and geotourism.

**Daniel H. Olsen** is an Assistant Professor at Brandon University, Manitoba, Canada, in the Department of Geography. His research interests include the geography of religion, religious and heritage tourism, tourism and disabilities, and tourism in peripheral destinations.

**Kate Rodger** is a Research Fellow in the School of Environmental Science at Murdoch University. Her research interests include human–wildlife interactions in the marine and terrestrial environment, identifying, and minimizing visitor impacts through visitor management techniques in protected areas, improving links between science and policy, and integrating ecological and social sciences in nature tourism research.

**Eric J. Shelton** teaches tourism and is interested in the relationship between conservation and tourism while he seamlessly combines an academic career and environmental restoration activity as a trustee on an ENGO. In particular, he views the urban setting as a promising location for the production of biodiversity. Being based in Dunedin, New Zealand offers him opportunities to work toward producing a nature-based tourist city.

**Mick Smith** is Associate Professor and Queen's National Scholar jointly appointed to the Department of Philosophy and the School of Environmental Studies at Queen's University, Kingston, Ontario. He is author of *Against Ecological Sovereignty: Ethics, Politics and Saving the Natural World* (2011). He is also editor of the Elsevier journal *Emotion, Space and Society*.

**Robert Steiger** is a Postdoctoral Fellow at the Institute of Geography, University of Innsbruck, Austria. He is also senior scientist at alpS-Centre for Climate Change Adaptation Technologies, Innsbruck, Austria.

**Emma J. Stewart** is a Senior Lecturer in Parks and Tourism at Lincoln University, Christchurch, New Zealand and a Research Associate at the Arctic Institute of North America, University of Calgary, Canada. She has a long-standing interest in polar tourism.

**Johann Stötter** is Professor for Geography at the University of Innsbruck (Austria). He is also Scientific Director of the alpS-Centre for Climate Change Adaptation Technologies and Deputy Director of the Institute for Mountain Research: Man and Environment of the Austrian Academy of Sciences.

**Arvid Viken** is a Professor in Tourism at University of Tromsø, Norway. His major focus is on tourism as a social, cultural, and spatial phenomenon, and on destination development. He has published several articles on tourism in the Arctic, particularly concerning Svalbard and in the Sami areas.

**Nicki Youroukos** is a Graduate Candidate in the Master of Environmental Studies, Nature-based Recreation and Tourism program at Lakehead University. Her research interests include rural tourism in Northwestern Ontario and small and medium enterprise (SME) owners.

# Foreword

In 2003, I was part of an international group of media invited to the Three Gorges region of China's Yangtze River. The occasion was the completion of a major phase in the construction of the monumental Three Gorges hydroelectric plant – the largest such project in world history. For many of us, it was also a final opportunity to experience a region that would be changed forever.

While Chinese government officials insisted that the dam was important to save lives lost each year as a result of flooding, a byproduct of the project was a dramatic rise in the river's water level. As a result, sections of the Yangtze River, including where it flows through the narrow Three Gorges, would be less dramatic. Entire towns that had existed for centuries were to be lost to the rising waters and, with it, a way of life. In short, this trip represented the last chance to experience one of the most unique places on earth before it was changed forever.

The damming of the Yangtze River, and the resulting change to the region's landscape, ecosystem, and culture, drew considerable worldwide attention. To many, it became a symbol for how humanity was altering and, in some cases, destroying natural wonders all over the world. Regardless of where one stands on the practical necessity of the Three Gorges dam, there is no disputing that its creation ended a chapter in the region's long history, and that future generations will not have the chance to experience the destination in the same way that others had before. Visiting the Three Gorges might still be a magnificent travel experience but, without a doubt, something has been lost.

To me, this experience gets to the essence of last chance tourism. The desire on the part of travelers to experience destinations motivated by the knowledge that these places are on the verge of changing, or disappearing, forever. These changes can be the result of climate change, or due to the encroachment of civilization.

As the Editor-in-Chief of *TravelAge West*, a U.S. travel industry trade publication my career is spent watching travel trends and last chance tourism has been a trend I've been aware of for some time. In 2007, I wrote one of the first editorials about this phenomenon calling it, "the tourism of doom" (Shapiro 2007). In the article, I wrote about a family member who had decided to travel to Patagonia motivated by the desire to experience the region before climate change altered it. At the same time, I met other travelers who had similar motivations for traveling to destinations such as Greenland, the Andes, Mount Kilimanjaro, the Galapagos, the Maldives, and more. Not long afterward, my editorial became the focus of a story in *The New York Times* (Salkin 2007a, 2007b) and the global impact of this article was surprising even to me. Soon I was being interviewed by media outlets all over the world, from National Public Radio in the U.S. to the Australian Broadcasting Company to newspapers in South America. It was clear that last chance tourism, as it has come to be called, hit a nerve with people everywhere.

As the various contributions in this edited volume highlight, tour operators and suppliers throughout the travel industry have responded to this consumer demand by adding itineraries and promoting new destinations with the clear knowledge and intent of capitalizing on last chance tourism. Previously unknown destinations such as Churchill, Canada, suddenly burst onto the tourism map because of its easy accessibility to polar bears – a vulnerable species. Cruises to Antarctica are sold out far in advance with tourists who want to see the unspoiled landscape of the Southern Land before it "disappears." Without a doubt, last chance tourism has claimed a foothold in the travel industry.

The rise of last chance tourism has also coincided with other travel attitudes. In particular, one of the buzz words of the travel industry over the past few years has been "authentic." More and more travelers are expressing the desire to experience places in authentic ways that are not packaged for them by marketing experts. Baby boomers – the traveler demographic that is most influential in the industry at the moment – are particularly adamant about experiencing destinations in a way that is local, authentic, and unspoiled. Interestingly, they share these values with the youngest generation of travelers. At its root, last chance tourism is about those places that are still authentic. Places that, at least for the time being, remain as they have been for centuries, if not longer. Ultimately, the window of opportunity for a local, authentic and unspoiled experience of many of these places is closing, and travelers are responding to that urgency.

The future of last chance tourism appears strong. While the world continues to flounder piecemeal in its response to the threats to these fragile places, factors such as transportation innovations, technology, and population growth mean there will be ever more travelers seeking out these destinations even as they vanish before our eyes. The expected impact of an emerging middle class in countries such as China, India, and Brazil that participates in more international travel, alone has the potential to vastly change the global tourism landscape. Unless steps are taken to protect some of these threatened areas, the devastation seems unavoidable.

Still, there are some signs of change. It is much more common these days for a hotelier, for example, to acknowledge that it is in his or her best interest – as a businessperson as well as a human being – to protect the beach and the ocean that is the lifeblood of their resort. Throughout the travel industry, there is a growing understanding that if we allow the beloved sights of our planet to be destroyed for short-term gain, we will be killing off the very essence of why people travel in the first place.

In discussing last chance tourism, another common concern is often voiced, namely that by rushing to see these fragile destinations, tourists are hastening their demise. In other words, by calling attention to these places, we run the risk of doing further damage – we are loving them to death. Certainly, this is a valid consideration but, from what I've witnessed on my travels and in my discussions with fellow travelers, I strongly feel that in this case, the rewards greatly outweigh the risks.

In so many ways, the experience of travel emotionally connects individuals with distant people and places in a much more meaningful way than is possible otherwise. In a situation that requires a substantial shift in mindset in order to save many of these endangered environments, as is iterated in the conclusion of this volume, I believe one of the greatest tools we have is to expose people to the wonders that are on the verge of disappearing. In the same way that someone who has experienced the fragile and miraculous wildlife in Africa, for instance, is far more likely to come home a conservationist, we need to show people the consequences of maintaining the status quo – the magnitude of what is at stake. This is not to say, however, that in all cases we should continue with the same free-wheeling, market-driven tourism approach currently in place. But a carefully examined strategy, with moderated controls and, in some cases, strong restrictions, will not only protect these destinations far more than what we are currently seeing, but it will also allow these places to do what they do best – inspire visitors. You can see examples of this approach discussed throughout this book including many ways in which it has been a success. Finally, the most important aspect of last chance tourism is for those of us within the tourism industry as well as travelers to fully embrace the concept that our natural world is changing, and that many of our most prized destinations are seriously at risk. Working together, the travel industry is in the best position to appeal to governments to take measures to slow down and possibly even reverse some of the destruction we're seeing.

Kenneth Shapiro
July 2011

## References

Salkin, A. (2007a, December 17). "Tourism of doom on the rise." *New York Times*. www.nytimes.com/2007/12/16/world/americas/16iht-tourism.1.8762449.html

Salkin, A. (2007b, December 17). Before it disappears. *New York Times*. www.nytimes.com/2007/12/16/fashion/16disappear.html

Shapiro, K (2001, May 11). The tourism of doom. *TravelAge West*. www.travelagewest.com/Travel-News/Viewpoint/The-Tourism-of-Doom/

# Acknowledgments

Foremost, we would like to thank of all 23 authors who contributed to this book for their support and contribution. The dedication and willingness to address a relatively new topic in tourism is commendable.

We are extremely grateful for the support and encouragement provided by Routledge. In particular, Emma Travis provided great foresight at the outset of the project, and Faye Leerink, the editorial assistant, and the editorial team have patiently supported us throughout the process. It has been a pleasure working with such a professional and friendly team!

We would also like to thank our colleagues, funding organizations such as the Social Sciences and Humanities Research Council of Canada, and friends who have provided their support. Finally a special thanks to our families, who despite having endured earthquakes, relocations, promotions, and various rants, have persevered and continued to provide their unconditional love and support throughout this endeavor.

# Section I

# Disappearing landscapes/seascapes and vanishing fauna

# 1 An introduction to last chance tourism

*Raynald Harvey Lemelin, Emma Stewart, and Jackie Dawson*

The last chance to consume, collect, or photograph rare or exotic specimens is not a new phenomenon in leisure and tourism. Indeed, as Bowman and Pezzulo (2010) and Urry (2004) noted, many of the earliest tourist destinations were 'often considered "places to die for" in the sense that one "must see" them before one dies' (Bowman and Pezzulo 2010: 191). Places, contends Urry (2004), also die, 'insofar as they either are exhausted in terms of their capacity to offer the experiences today's tourists want or else they become inhospitable to life due to earthquakes, volcanic activity, pollution, hurricanes, decay, and so forth' (Urry cited in Bowman and Pezzulo 2010: 190). Yet, unlike these past experiences which were often confined to one species, one site, or one geographical region, last chance tourism provides the opportunity to witness the demise of ecosystems, to behold the extinction of an entire species from its natural habitat. Tourism of this nature is for all intents and purposes a chance to observe ecocide first-hand. Critics have also labelled last chance tourism as a short-term marketing ploy aimed at selling more guidebooks or increasing profits from destinations in potential demise. The ethics and motivations underlying this form of travel is what makes last chance tourism so interesting, particularly at a time when global climate-induced anthropogenic changes are proceeding at an alarming rate. After all, who would want to be the last individual to witness the tumble of the final glacier in Antarctica or Greenland; to observe the last breath of an emaciated polar bear in Churchill, Canada; to step on the last ice of Mt. Kilimanjaro? According to various researchers, including a number of authors featured in this book, many tourists concerned with vanishing destinations such as the Great Barrier Reef or disappearing wildlife such as mountain gorillas and polar bears are purchasing these last chance experiences (Agnew and Viner 2001; Amos 2001; Becken and Hay 2007; Hall and Higham 2005). As Nuttall (2010) noted during one of his excursions to Antarctica, the need to participate in last chance tourism should come as no surprise since visitors 'are made to feel that they are travelling in an environment that is itself at risk. . . . under threat of disappearing, as the ice melts and yet another ice shelf collapses' (p. 210). In other instances, travellers are advised to visit such national parks as Glacier, Joshua and Biscayne and Virgin Island in the USA, before the glaciers, the trees, and the coral reefs are gone (Shapley 2011).

Since the 1990s, the mainstream literature, and increasingly travel blogs and other forms of modern media, have alerted the public to last chance issues. More recently, a special issue of *Newsweek* (2010) listed 100 places 'to remember before they disappear', the book 100 Places to Remember Before they Disappear (Drew 2011) and the accompanying website of the same name, 'features 100 photographs from one hundred different places around the world at risk of disappearing' (100 Places.Com, 2011). Perhaps the most influential text was Douglas Adams and Mark Carwardine's (1991) *Last Chance to See* and later, Carwadine's (2009) *Last chance to see: in the footsteps of Douglas Adams.* The books inspired a BBC television production and a specialized tour entitled 'Last Chance to See Collection' offered by Bales Worldwide in 2009 (Easier Inc. 2011). Other books include: Addison's (2008) *Disappearing world: 101 of the Earth's most extraordinary and endangered places*; Garbutt's (2007) *100 Animals to See Before they Die*; Lisagor and Hansen's (2008) *Disappearing Destinations: 37 places in peril and what can be done to help save them*; and Travis's (2009) *Frommer's 500 Places to See Before They Disappear.*

The outcome from this attention is an emerging phenomenon in tourism known as last chance tourism (Dawson *et al.* 2010; Lemelin *et al.* 2010). Last chance tourism can be defined as when 'tourists explicitly seek vanishing landscapes or seascapes, and/or disappearing natural and/or social heritage' (Lemelin *et al.* 2010: 478). This includes the desire to observe, photograph and interact with animals that may be endangered, threatened or rare, and are increasingly being offered by operators in a number of destinations around the world (Ballantyne *et al.* 2009; Garbutt 2007; Newsome, Dowling and Moore 2005; Munn 1992). Similar to climate change tourism (Becken and Hay 2007; Hall *et al.* 2011; Jones and Phillips 2011; The World Tourism Organization and the United Nations Environment Programme 2008), last chance tourism is often stimulated by the perception of change brought about through climate change, but unlike climate change tourism, last chance tourism is also about visiting destinations that may be in decline due to other anthropogenic factors, or even in some cases, visiting cultural sites, where decline may have nothing to do with climate change (i.e., globalization, modernization). By providing opportunities to see the living dead, or witness the demise of places, species, and people, last chance tourism can also be defined as the precursor to dark tourism (Lennon and Foley 2000; Seaton 2002). If dark tourism is the embodiment of loss, last chance tourism is not only the epithet, but the call to action. Indeed, it is an opportunity to learn and possibly reverse these changes before it is too late. It is a chance to enlighten the dark.

## Perceptions of rarity

As various authors in this book allude, perceptions of rarity are fuelled and at times even manipulated by various interest groups including the scientific community, marketers, travel operators, researchers and journalists. Publicizing the vulnerability of certain threatened wildlife or destinations through wildlife documentaries or fundraising campaigns can be as Burns and Bibbins (2009) and Dawson *et al.* (2010) argue a 'call to action' to protect and preserve these endangered destinations. In the case of wildlife tourism, seeing an endangered species in its natural habitat brings an emotional bond that is rarely acquired in a zoo (Garbutt 2007). Through direct in-situ experience, Garbutt (2007) suggests that visitors can develop an emotional bond, where individuals learn and want to help by becoming last chance to see ambassadors. The call to action can also result as a 'call to visit' these destinations, so to speak; to see them 'before' or 'in case' they disappear.

Courchamp *et al.* (2006) hypothesized that if consumers place disproportionate value on rare species, this could result in a cycle in which increased exploitation further reduces the population size, which in turn increases its value and ultimately leads to its extinction in the wild – a process they termed the 'anthropogenic Allee effect' (Hall *et al.* 2008: 75). The anthropogenic Allee effect and the extinction vortex is explained in the following fashion. When certain species acquire a rarity value, prices for scarce species can increase significantly even though continued exploitation can compromise the welfare of the attraction (Hall *et al.* 2006). Researchers call this latter occurrence a negative correlation, when the rarity of a product actually increases the demand for the said product (Vaillant 2010).

As the value of a rare item increases, more time, effort, or resources may be devoted to trying to acquire it, increasing the pressure on the species as it becomes rarer. The idea is that as long as someone will pay any price for the rarest of the rare, market price will cover (and exceed) the cost of harvesting the last specimen (Courchamp *et al.* 2006). Perceived rarity when associated to other anthropogenic stressors such as climate change, pollution or increasing visitation demands can result in the rapid demise or local extinction of certain species (Hall *et al.* 2008). Further complicating the protection of these features are cultural traditions, and how the acquisition, consumption, and/or ownership of these rare products conveys social prestige and status (e.g., the species bucket list). Indeed, listing in any form can be perceived as official verification that the area in question is a limited edition (Rivalan *et al.* 2007), as such, these lists are well suited to appeal to those travellers who 'collect places'. In this case, the visitor is no more than the final consumer of goods to which they are delivered 'just in time' before that destination's fate is sealed.

However, it does not have to be this way, since last chance tourism is not just a search for animals on the edge of extinction or for disappearing destinations or human populations threatened by acculturation, it can also be a celebration of conservationism. For across the world, thousands of individuals, including tourists, dedicate their time, often their lives in some of the harshest places on earth, in an effort to help species or rehabilitate habitats from the brink of extinction (Green 2008). In the case of the kakapo, orang-utan and mountain apes, these efforts have saved the animals from the brink of extinction in the wild. In his book *Frommer's 500 Places to See Before They Disappear*, Travis (2009) suggested that indeed many positives can come from these last chance tourism experiences.

I discovered more positive developments than I expected. Some of the destinations I have included in this book are already on the road to being saved, usually because they have been championed by preservationists with a will to make a difference. And even in cases where a site has been lost, or irrevocably damaged, it often has become a rallying point for activists, inspiring them to fight on so that the same mistakes are not made again. Many of the case studies in this book are reasons for hope, not despair, and the more support we can lend them, the better (Travis 2009: vi).

From this perspective, last chance tourism pays homage to biodiversity and conservationism, and in general, 'seeks to oppose the reduction of an environment (or a culture) to a novelty, curiosity, object or potential resource. Indeed, "last chance" tourists often carry a residual hope that they will not, after all, be the "last to see" and that some things might yet miraculously survive' (Smith this volume). Instead of focusing on the negative, we want to inspire a message of hope in each chapter, whether this is through a better understanding of the visitors or the on-site operators or through regulatory mechanisms of some kind. Where conservation has embraced neoliberal global capitalism, then last chance to see is morbid; when last chance tourism plays a role in a rescue from the brink of extinction – it is inspiring.

## Book overview

Through a multi-authored, multi-disciplinary volume featuring researchers in the fields of leisure, tourism, anthropology, geography and sociology this book attempts to address the nuances of managing sites in the Arctic and Antarctic, Aoteora/New Zealand, Australia and China, and various animals such as mountain gorillas, polar bears, albatrosses and kakapos. Several chapters document studies attempting to understand visitor behaviour and attitudes, and in some cases responses to last chance tourism. Other chapters also remind us that last chance tourism is not simply relegated to ecological spheres; last chance tourism has had and will continue to impact local communities, and will demand adaptation at multiple scales.

The book is divided into three sections each with its focus and particular descriptions of the last chance tourism phenomenon. Section I, 'Disappearing Landscapes/Seascapes and Vanishing Fauna', provides an international overview of changing landscapes and seascapes, and threatened species and includes chapters that examine the physical aspects of last chance tourism in the Arctic, Antarctic and Alpine environments. Some of the most notable findings in these three chapters are the following. Johnston *et al.* suggest that when something disappears, something else appears, thus last chance tourism should be linked to *first chance tourism*. The role of climate-change ambassadors, according to Lamers *et al.*, in Antarctic tourism is limited, yet other perhaps more effective interpretation strategies do exist. Contesting the notion of disappearing landscapes is Steiger *et al.* who report that while some skiing sites in the US and Europe will disappear, some will remain, and in fact are likely to thrive. Some of the most popular known species including orang-utans and mountain gorillas, various bird species and polar bears are discussed in the chapters by Newsome and Rodger, Hvenegaard and Stewart *et al.*

Section II, 'Media and Marketing', provides some of the earliest documentation of marketing and media research examining last chance tourism through case studies that focus on destinations in North America, Greenland, Africa and Australia, including the Great Barrier Reef, Mt Kilimanjaro and the Ilulissat Icefjord. Highlights from this section include the disconnect between research interests (i.e., scientists, conservation groups), media sources and local operators noted by Olsen *et al.* and Coghlan. A similar conclusion was reached by Frew who noted that although many operators highlight the vulnerability of certain destinations, few are actually using the concept of last chance tourism to specifically market their products.

Section III, 'Ethics, Culture and Management', discusses the social, ethical and political dimensions of these vanishing destinations in China, Northern Canada, New Zealand and, more specifically, national parks in Canada. These are some of the earliest discussions on the socio-cultural and political dimensions of last chance tourism. The legacy of colonialism in our approach to nature and other societies are discussed by Smith, and Lemelin and Baikie, while Shelton, and Lemieux and Eagles highlight how national initiatives can help to bring about change and conserve endangered species and preserve national parks in an era of global change.

The conclusion draws together the key themes and lessons that can be synthesized from the contributions in this volume and outlines various forms of management solutions that are currently being implemented to deal with last chance tourism.

In many ways, like Douglas Adams and Mark Carwardine, we want to celebrate conservationism and resiliency. Like these authors, we would like to re-examine these case studies in 20 years from now, and hope to use the famous quote from the poet and playwright Mark Twain, that the news of the demise of these areas and fauna, "was an exaggeration".

## References

Adams, D. and Carwardine, M. (1992) *Last chance to see*. New York: Ballantine Books.

Addison, A.C. (2008) *Disappearing world: 101 of the Earth's most extraordinary and endangered places*. New York: Harper Collins.

Agnew, M.D. and Viner, D. (2001) 'Potential impacts of climate change on international tourism', *Tourism and Hospitality Research* 3(1): 37–60.

Amos, J. (2001) 'Kilimanjaro's white peak to disappear', *BBC News from San Francisco*. Online: news.bbc.co.uk/1/hi/in_depth/sci_tech/2001/san_francisco/1177879.stm (accessed 26 April 2011)

Ballantyne, R., Packer, J. and Hughes, K. (2009) 'Tourists' support for conservation messages and sustainable management practices in wildlife tourism experiences', *Tourism Management* 30: 658–664.

Becken, S. and Hay, J. (2007) *Tourism and climate change: risks and opportunities*. Clevedon, UK, and Buffalo, NY: Channel View Publications.

Bowman, M.S. and Pezzulo, P.C. (2010) 'What's so 'dark' about 'dark tourism'? Death, tours, and performance', *Tourist Studies* 9(3): 187–202.

Burns, P. and Bibbins, L. (2009) 'The end of tourism? Climate change and societal challenges. Twenty-First Century Society', *Journal of the Academy of Social Sciences* 4(1): 31–51.

Carwardine, M. (2009) *Last chance to see: In the footsteps of Douglas Adams*. London: HarperCollins Publishers.

Courchamp, F., Angulo, E., Rivalan, P., Hall, R.J., Signoret, L., Bull, L. and Meinard, Y. (2006) 'Rarity value and species extinction: the anthropogenic allee effect', *PLoS Biology*, 3(12): 2405–2410.

Dawson, J., Stewart, E.J., Lemelin, H. and Scott, D. (2010) 'The carbon cost of polar bear viewing in Churchill, Canada', *Journal of Sustainable Tourism*, 18: 319–336.

Drew, P. (2011) *100 Places to remember before they disappear*. Harry N. Abrams.

Easier Inc. (2011) 'Tour operator launches "Last Chance To See" itineraries.' Online: www.easier.com/55858-tour-operator-launches-last-chance-to-see-itineraries.html (accessed 26 April 2011).

Garbutt, N. (2007) *100 Animals to see before they die*. Bradt Travel Guides.

Green, T. (2008) 'The thing about extinction'. Online: www.bbc.co.uk/lastchancetosee/sites/about/extinction.shtml (accessed 26 April 2011).

Hall, C.M. and Higham, J. (2005) *Tourism, recreation and climate change*. Clevedon, UK, and Buffalo, NY: Channel View Publications.

Hall, C.M., Gossling, S. and Scott. D. (2011) *Tourism and climate change: impacts, adaptation & mitigation*. New York: Routledge.

Hall, R.J., Milner-Gulland, E.J. and Courchamp, F. (2008) 'Endangering the endangered: the effects of perceived rarity on species exploitation', *Conservation Letters* 1: 75–81.

Jones, A.L. and Phillips, M. (2011) *Disappearing destinations: climate change and the future challenges for coastal tourism*. CABI.

Lemelin, H., Dawson, J., Stewart, E.J., Maher, P. and Lück, M (2010) 'Last chance tourism: the doom, the gloom, and the boom of visiting vanishing destinations', *Current Issues in Tourism* 13(5): 477–493.

Lennon, J. and Foley, M. (2000) *Dark tourism: The attraction of death and disaster*. London: Continuum.

Lisagor, K. and Hansen, H. (2008) *Disappearing destinations: 37 places in peril and what can be done to help save them*. New York: Vintage Departures.

Munn, C.A. (1992) 'Macaw biology and ecotourism, or "when a bird in the bush is worth two in the hand"', in S.R. Beisssinger and N.F.R. Snyder (eds.), *New World Parrots in Crisis: Solutions from Conservation Biology*, Washington, DC: Smithsonian Institution Press, pp. 47–72.

Newsome, D., Dowling, R.K. and Moore, S.A. (2005) *Wildlife Tourism*. Clevedon, UK, and Buffalo, NY: Channel View Publications.

*Newsweek* (2010) '100 places to remember before they disappear'. Online: www. newsweek.com/feature/2010/100-places-to-remember.html (accessed 3 April 2011).

100Places.Com. (2011) 'Welcome to 100Places.Com'. Online: www.100places.com/en/WELCOME TO 100PLACES.COM (accessed 28 April 2011).

Nuttall, M. (2010) 'Narratives of history, environment and global change: expeditioner-tourists in Antarctica', in C.M. Hall and J. Saarinen (eds.), *Tourism and change in polar regions: climate, environment and experience,* Abingdon: Routledge.

Rivalan, P., Delmas V., Angulo E. *et al.* (2007) 'Can bans stimulate wildlife trade?' *Nature* 447: 529–530.

Seaton, A.V. (2002. 'Thanatourism's final frontiers? Visits to cemeteries, churchyards and funerary sites as sacred and secular pilgrimage', *Tourism Recreation Research* 27: 73–82.

Shapley, D. (2011) '87 Endangered national parks to see now', *thedailygreen*. Online:

www.thedailygreen.com/living-green/threatened-national-parks-47091804 (accessed 28 March 2011).

Travis, W. (ed.) (2009) *Frommer's 500 places to see before they disappear*, Hoboken, NJ: Wiley Publishing, Inc.

The World Tourism Organization and the United Nations Environment Programme (2008) *Climate change and tourism: Responding to global challenges.* Madrid: The World Tourism Organization.

Urry, J. (2004) 'Death in Venice', in M. Sheller and J. Urry (eds.), *Tourism mobilities: places to play, places in play.* London: Routledge, pp. 205–215.

Vaillant, J. (2010) *The tiger a true story of vengeance and survival.* Toronto: Knopf Canada.

# 2 Firsts and lasts in Arctic tourism

## Last chance tourism and the dialectic of change

*Margaret Johnston, Arvid Viken, and Jackie Dawson*

### Introduction

This chapter examines the concept and reality of last chance tourism (LCT) in the context of Arctic environments. The Arctic is a vast biophysical region that is composed of many sub-regions: areas that maintain similarities to the larger region, but display particular sub-regional characteristics. These sub-regional variations are significant in both tourism and climate change. The biophysical Arctic is a cold ocean with substantial ice coverage surrounded by land (Bone 1992; Chaturvedi 1996; Sugden 1989). The Arctic also has culturally constructed definitions that reflect the images and meanings that people hold of the region (Jacobsen 1994). These are particularly relevant for tourism given its reliance on image and meaning, and in the Arctic many of these views reflect the natural assets of landscapes, wildlife and climate (see Saarinen and Tervo 2006; Hall, Müller and Saarinen 2009; Lundmark 2010).

Tourism makes use of a diverse range of biological, physical and cultural resources that vary in important ways across the Arctic. Thus, the dramatic changes being experienced in parts of the Arctic, and those which are predicted for the region, have tremendous potential to influence tourism because of this heavy reliance on the environment and its particular components as attractions. Global environmental change, based on a changing climate system and subsequent series of outcomes that affect the landscape and local people, has led to issues requiring resolution as well as opportunities for the tourism industry, for communities and for tourists.

The popular press has emphasized the extreme vulnerability of Arctic tourism destinations, but academic interest in LCT in the region has been limited. However, the topic has been explored to some degree by several researchers focused on polar regions (Lemelin and Johnston 2008; Dawson, Lemelin and Stewart 2009; Dawson, Stewart, Lemelin and Scott 2010; Lemelin, Dawson, Stewart, Maher and Lück 2010; Eijgelaar, Thaper and Peeters 2010). This emerging literature identifies last chance motivation as a growing theme in polar tourism (Dawson *et al.* 2009, 2010; Eijgelaar *et al.* 2010) and demonstrates the use of the last chance theme in marketing (Frew 2008; Lemelin *et al.* 2010).

This chapter seeks to expand upon our understanding of LCT in Arctic environments by outlining the multi-scale changes arising from climate change, examining management responses to change, and describing responses of visitors and the industry, considering the marketing of LCT in the Arctic, and reviewing the concept through a critical lens via an exploratory study and brief examination of two instructive situations. The chapter examines LCT as a biophysical experience based on evidenced or predicted changes to landscape and wildlife and also as a cultural experience, reflecting perceived changes in identifiable cultural commodities. Through this discussion, the chapter considers the concept and reality of LCT in order to come to some understanding of change in Arctic tourism. We begin by discussing the general context of LCT in the Arctic through a brief introduction to the experienced and predicted outcomes of climate change and through a discussion of the responses of the public, the industry, communities, tourists, and managers.

## Last chances in one of the last frontiers

The Arctic is particularly vulnerable to climate change and related environmental changes. Several major changes have already been experienced: increasing average summer and winter temperatures, decreasing extent of sea ice in summer, altered precipitation patterns, and changes in population distributions of some species (ACIA 2004). The Arctic Climate Impact Assessment (ACIA 2004) was the first comprehensive examination of likely outcomes of climate change for the Arctic. Based on scientific, indigenous, and local observations and knowledge, this report describes expected impacts for the Arctic and its sub-regions through the use of a series of scenarios given a confidence rating by climate experts. The report uses ten themes to summarize outcomes and predictions, and these themes include all aspects of environmental and human use. The ACIA has been followed by several publications that identify observed changes and make further predictions for the Arctic and its sub-regions including 'State of the Arctic', a report produced by the National Oceanic and Atmospheric Administration (NOAA 2006) and 'Arctic Climate Impact Science – an update since ACIA' produced by the World Wildlife Fund (WWF 2008).

Several tourism researchers have considered the pan-Arctic and/or regional implications for tourism of observed and predicted changes, especially changes in sea ice cover and extent, distribution and health of wildlife populations, precipitation patterns, and temperature patterns (e.g. Johnston 2006; Dawson, Maher and Slocombe 2007; Hall *et al.* 2009; Lundmark 2010). In addition to these general explorations of outcomes, various authors have sought to aid in the further understanding of specific climate-induced changes and the implications for tourism sectors or activities. For example, Stewart *et al.* (2007) consider ice regimes in the Canadian Arctic and the implications of sea ice variability for cruise tourism in terms of both access to sites and communities and the availability of attractions such as ice-dependent wildlife. Dawson *et al.* (2010) examine climate-related change, sustainability and adaptation issues in the polar bear viewing industry in Churchill, Manitoba, Canada and note that the polar bear has become the most evocative species in the media showing the struggle to survive in a changing environment.

Public awareness of climate change appears to be growing throughout the world as the scientific community, governments and the media continue to make information available about global and local outcomes. The concept of LCT may well have emerged first in relation to publicity about the changes being experienced in and predicted for the Arctic, especially as the general public became aware of melting glaciers, changing ice distribution and the health of polar bear populations (University of Calgary 2009). This position is reinforced by Mintel International Group Ltd (2008, cited in Eijgelaar *et al.* 2010, 338): 'Tour operators report that more and more travelers are asking about trips to the Arctic, evidently believing that it might vanish at any minute. They want to get there before the ice cap melts and the animals – especially the polar bears – drown or disappear.'

The surge of climate-related Arctic tourism promotion and possibly tourist interest in actual travel has been described by Lemelin *et al.* (2010), and Hall and Saarinen (2010a, 2010b). Lemelin *et al.* (2010) describe the developing demand for tourism opportunities in the vulnerable Arctic ecosystems that some tourism operators have begun to see and promote. For example, the decreased extent of sea ice cover appears to have played a role in an observed increase of cruise ship activity in Arctic Canada, with a doubling of trips between 2005 and 2006 (Buhasz 2006) and a 70 per cent rise in cruise itineraries through the increasingly navigable Northwest Passage between 2007 and 2010 (Stewart, Draper and Dawson 2010). Pagnan (2003) reports that tourism professionals in Alaska had observed changes in wildlife and in the landscape that have important repercussions for the tourism industry, including the extension of the season for cruise tourism along the southwest coast. Similarly, residents of the sub-Arctic community of Churchill, Manitoba report a lengthened tourism season for polar bear viewing due to the increasingly ice-free conditions of Hudson Bay, which essentially forces polar bears to wait along the shorelines for longer periods of time until the ice freezes (Stewart 2009).

If tourists do have an awareness of climate change issues in the destinations they choose, this might mean they are more likely to attribute differences between their expectations and their experiences to the effects of climate change. Pagnan (2003) describes concerns of Greenland tourism officials, who noted that visitors to the west coast were disappointed with a lack of ice and ice-dependent wildlife, a response echoed by cruise tourists in the Canadian Arctic (Maher and Meade 2008; Maher 2010). Saarinen and Tervo (2006) found that the perception of a changing climate was seen by entrepreneurs as causing some visitors to choose to travel to Northern Finland rather than to the south because of the belief that the south was lacking snow. Tourists were seen as responding to the negative and positive changes they experienced or expected. Dawson *et al.* (2010) extend this idea based on a study of visitors to Churchill in 2007; they predict behavioural adaptations that would involve an even greater increase in demand for polar bear viewing in response to tourists' recognition of climate change-induced vulnerability. They propose the development of an inverse relationship between the supply of healthy polar bears and the demand for polar bear viewing opportunities; a relationship that they consider will become a common response to impacts of climate change on vulnerable Arctic resources. Indeed, Dawson *et al.* (2010) described the responses of a number of tourists interviewed in Churchill in recent years who were motivated in large part by last chance tourism (also see Stewart *et al.* this volume). For example, one tourist stated: 'I thought I better come see the bears because the next time I am in this country they will be all gone' and another stated 'I wanted to see the bears with my daughter because my grandchildren and their children may not ever know polar bears except in a zoo.' Although last chance motivation certainly appears to be present among polar bear viewing tourists in Churchill, Canada the extent to which polar bear viewing tourists are motivated solely due to last chances remains unknown (Dawson *et al.* 2009; Lemelin and Wiersma 2007, Stewart 2009; Lemelin *et al.* 2010).

Community members in tourism areas and industry stakeholders likewise are responding to their perceptions about outcomes of climate change. For example, residents of Churchill identified climate change as one of the primary threats to tourism development in the community (Stewart and Draper 2007). In their industry study, Saarinen and Tervo (2006) reported that about half of the tourism entrepreneurs they interviewed in Finland clearly did recognize climate change as a problem, but the other entrepreneurs were sceptical about its reality and did not believe it would influence their businesses or the industry. In a follow-up study conducted two years later in 2007, Saarinen and Tervo (2010) found that the majority of tourism entrepreneurs surveyed believed in the reality of climate change and its impacts on the tourism industry. Further, some entrepreneurs had observed changes that affected their operations (e.g. having to temporarily shut down the business because of warm winter weather), and most were pessimistic about the future of winter tourism in Finland, even though some gains in Northern Finland had been made because of poor snow conditions in European alpine destinations. Adaptation methods used by entrepreneurs in Northern Finland included using snow-making facilities, careful marketing, and contingency planning, methods used in not only the skiing sector, but also other snow-based sectors (Tervo-Kankare and Saarinen 2011).

Environmental changes bring difficulties as well as opportunities for the tourism industry and the communities that rely to any degree on tourist visitation. As communities, the industry and other tourism stakeholders, and tourists themselves adapt to changes, the possibility of LCT has been identified as an explicit theme, reflecting the motivation of tourists to see particular wildlife attractions or landscapes before they are irrevocably altered. In tandem with the development of this tourist interest is the explicit marketing of LCT by operators who seek to capitalize on that motivation – or who might be trying to foster that motivation. LCT has been defined as a niche tourism market where tourists seek out diminishing destinations or those with features that are disappearing (Lemelin *et al.* 2010). This type of tourism includes natural landscapes, biota and social and cultural components of a place (Lemelin *et al.* 2010). Although the foundation of LCT is based upon the explicit motivation amongst tourists to travel to vulnerable destinations and view vulnerable phenomena, the extent to which individuals are solely motivated or even motivated in part by this perceived vulnerability remains unclear (see Lamers *et al.* this volume).

Given that the Arctic is among the destinations profiled in the popular press as being most affected by climate change, it is a good candidate for LCT marketing; though for the most part, tourist operators and other stakeholders have not emphasized this in an overwhelming way (Frew 2008). For example, in Churchill, an obvious location for such marketing to be used given its reliance on polar bear viewing tourism, operators do not appear to be particularly interested in categorizing their product in this way (see Ord 2010).

In what appears to be the only study thus far to examine the explicit marketing of the last chance theme in tourism, Frew (2008) found that several operators were marketing last chance tours in the Arctic (also see Frew, this volume). For example, in Greenland three different last chance options were offered and all had a climate change-related theme of an educational nature. Frew (2008) also found evidence of the commercialization of LCT that was aimed more at sensationalizing the product than promoting the educational aspect of LCT. Despite these few explicit examples, a very small number of operators in the study were actually using the last chance theme in their promotion. Frew (2008: 20) notes that

> particular sectors of the media . . . have been using aspects of global warming and disappearing natural attractions as interesting and exciting angles for their stories but, in reality, few tour operators are using this aspect to promote their tourism products.

## Rhetoric and realities – two sides to the last chance tourism industry

In order to explore current use of the LCT concept, an exploratory telephone survey was conducted in early 2011 by one of the authors of this chapter. Information was gathered from 12 individuals involved in Arctic tourism. Ten of these individuals were key stakeholders in the tourism industry in Svalbard and Northern Norway. These interviewees represent the most popular tourism products and destinations in the Norwegian Arctic, such as Hurtigruten, the cruise and tourism industry on Svalbard (see Figure 2.1), the Lofoten Islands, and North Cape, and also include a representative of the Norwegian international tourist marketing organization, and a tourism consultant. In addition, a northern travel journalist and a tour operator both with detailed knowledge and experience of Greenland tourism were interviewed. All interviews commenced with the question whether the individual knew of the term 'last chance tourism' and followed with a discussion about whether this type of climate change tourism might be happening. They were also asked whether and how the tourism industry was using aspects of the climate change discourse in their products and marketing.

This brief study of stakeholders specifically working in Arctic tourism in

*Figure 2.1* Tourists in Svalbard (Photo credit: Ole Magnus Rapp)

Greenland and Norway showed that none of the tourism providers were using the LCT term and further that none of them knew of any tour operators promoting their tours under this label. Further, the operators interviewed did not believe they have had many last chance tourists as clients. None of those interviewed in Norway believed that the climate situation was even critical, particularly since, as most noted, in the winter of 2011 there was more ice in the Svalbard archipelago than there had been for many years and the stock of polar bears and walruses continued to grow (though some noted a concern about the health of polar bears and the existence of toxins in the population as being a critical issue). In Greenland, despite observations of the recession of ice in the fjords, interviewees noted that the thickness of the ice cap is increasing, leading some to believe that there is no risk that the glaciers and icebergs will disappear in the near future. A tour operator in Lofoten, a well-known group of islands in Northern Norway, maintained that each époque experiences change and has its challenges: the issue now is climate change.

Despite the lack of acknowledgement of the LCT concept amongst operators in Greenland and Norway interviewed in this phone survey, several keys themes arise from the material covered in the interviews: the existence of new forms of tourism and the relationship of LCT to first chance tourism; and, the interaction of climate change, tourism change and other change. Each of these will be discussed briefly. First, there are examples of companies developing new tourism opportunities as a response to changes in the climate and at the landscape and/or seascapes levels. The minimum ice-edge moving northward and the rates of glacier calving in Greenland were some of the changes noted by participants. Since people in

Greenland are used to adapting to changing conditions, less ice in the fjords, resulted in a significant increase in boat-based tourism activities. Some of the interviewees specifically asserted that when something disappears, something else appears: thus in our understanding, LCT should be linked to first chance tourism. There are several examples of new tourism products that were proposed as possibilities arising out of climate change, beyond specifically climate change focused or last chance tourism. The example of new boating trips in the iceberg viewing areas of Greenland was mentioned, but others were suggested, such as paddling to the North Pole instead of skiing, new expedition cruise itineraries throughout the Arctic, also utilizing new modes of transport, such as yachts, and journeys around Spitsbergen, north of Greenland. With changes in the extent, thickness and distribution of sea ice, coastal sites and communities formerly inaccessible will become more accessible, especially for marine transportation.

Many of the development possibilities discussed in the interviews fit the desire in tourism for newness, otherness and exoticism (Wang 2000), and it is clear that actual changes in the environment open up the possibility of new tourism activities (see Aall and Høyer 2005). Further, there is the likelihood that developments reflect the spirit of adaptation and innovation that others have identified in the northern tourism world (e.g. Tervo-Kankare and Saarinen 2011). These new tourism forms illustrate the dialectic of tourism change: LCT (opportunities arising through vulnerability caused by changing conditions) is linked to first chance tourism (opportunities arising through new conditions).

A sub-theme in this dialectic is the interaction of research, management and tourism. Several of the interviewees noted the dual role of several of the actors in the climate change discourse. For example, the Norwegian Polar Institute was part of the Intergovernmental Panel on climate change (IPCC) and, in its role as a government adviser to Svalbard, along with the government authorities, has been responsible for a significant amount of tourism to Svalbard, comprising large numbers of ministers, ambassadors, journalists and environmental officers and specialists from all over the world. This climate change tourism activity did not exist to any great extent before the panel's report in 2004. Perhaps these are the real last chance tourists: visitors enjoying opportunities provided for them because of the new need for science, awareness and management.

Furthermore, numerous television programmes and websites have been produced focusing on climate and environmental changes, such as a programme showing the crown princesses and princes of the Scandinavian countries on a voyage providing them with knowledge of the Arctic environmental challenges. Such tours support the prevailing rhetoric of a vulnerable Arctic, giving argument for more research, research funding and protection. The same storyline has been observed in more tropical regions of the world including the highly publicized vulnerability of the Great Barrier Reef in Australia, which some have criticized for similarly capitalistic reasons (see Coughlin this volume).

Despite the scepticism about LCT revealed through the interviews, there are several examples of tour operators or cruise companies taking advantage of the climate change discourse and the fact that the Arctic has been labelled as a laboratory of, and showcase for, these changes. For instance, the Hurtigruten, a Norwegian coastal steamer company, arranged a climate-focused voyage to Svalbard in June 2010 (see Figure 2.2). The company invited several high-profile climate researchers on board as lecturers and the Norwegian Polar Institute was a collaborator, signalling the educational nature of the cruise. Such cruises are, of course, an example of the polar tourism paradox: their $CO_2$ factor is high, but they utilize a serious issue that engages customers to produce and market tourism opportunities, fully wrapping the product in politically correct rhetoric. In the tour promotion, it was maintained that:

> Svalbard is a showcase that illustrates the significant changes taking place in the Arctic; these are dramatic, and an indication of tendencies also valid for more southern areas. Together with outstanding representatives for the

*Figure 2.2* Cruising with Hurtigruten through the Norwegian Arctic (Photo credit: Ole Magnus Rapp)

science communities Hurtigruten has the pleasure to give her passengers an updated insight in the climate situation.

(Hurtigruten 2010).

It is clear that climate change interacts with other changes and together these can influence tourism. Several interviewees in the study described these interactions in Lofoten, North Norway, where climate change has had observable impacts, but so far these appear to have only indirectly affected tourism. The Lofoten Islands are known for their picturesque landscapes with steep mountains and cosy fishing villages, with accommodation in fisher huts (*rorbuer*). The sea surrounding Lofoten used to be the prime breeding area for the Arctic cod, and over the centuries a significant fishery and stockfish industry has developed. This fishery is about to disappear, as the migration and breeding patterns of the cod have changed. Due to higher sea temperatures in recent years, the breeding and thus the fishery activities increasingly have taken place in the districts north of Lofoten (Øseth 2010). However, this has not yet affected tourism, according to the interviewees, as it is the cultural landscape that has been the tourism attraction and because tourism and the fishery take place at differing times of the year. One of the interviewees suggested that the vanishing fisheries could make it easier to define Lofoten as a UNESCO World Heritage Site, for which the archipelago is currently a candidate. However, of relevance for this potential status and tourism in general is another threat: the potential for exploitation of oil and gas resources in the adjacent sea areas. Both local and national politicians want to exploit these resources. The plans are contested by most of the tourism industry, local fisheries and environmentalists, and the decision to proceed with the development has not yet been taken (as of early 2011). The pressure and uncertainty related to industrial resource development policies has created something of a passive approach to tourism development. There are two basic options: oil and gas, or tourism. If the planning decision is to pursue oil and gas exploitation, the idea has been discussed within the destination management organization circles to make use of the opportunity to market Lofoten as a LCT destination. However, the tourism stakeholders have had ethical concerns about pursuing this marketing strategy and the term has not been used thus far.

## Competition for access = no chance for tourism

These ideas expressed by the stakeholders who were interviewed are linked to another element in the competition for resources in the Arctic. It is not uncommon for tourism development anywhere in the world to be in competition for access to resources and it is typical of northern environments that this competition exists with subsistence, traditional, and rights-based access to resources. As seen above the situation can also be complicated by the need for access in the industrial development of fossil fuel and mineral resources. As climate change influences landscapes/seascapes, and flora and fauna, the competition for access to resources becomes increasingly more acute and thus requires more robust and, perhaps, detailed management. In some cases, the interaction of resource competition and the external stressors on the environment might lead not to LCT opportunities as described above for Lofoten, but rather to the exclusion of tourism before the last chance possibility is even recognized by the industry and by tourists, or is actively promoted. To illustrate the alternative to the first chance dialectic outcome, this discussion will now turn to an alternative outcome stemming from vulnerability by examining caribou hunting tourism in northern Canada. The complexity of competing access in the context of a changing climate is seen in the evolving management approaches to caribou populations in northern Canada and in the views on the reason for their seemingly atypical recent declines, which include the impacts of climate change and resource development, hunting, and natural population swings related to weather, disease, food and local conditions (Government of Northwest Territories 2006; Spitzer 2007; Nesbitt and Adamczewski 2009). Applications of climate predictions to factors of population dynamics indicates that the changing climate will result in herd declines for a variety of reasons, including an increase in weather variability (e.g. Gunn 1995; Brotton and Wall 1997). There are concerns across the circumpolar world about caribou declines; Canadian management responses have included in two places a ban on tourist sport-hunting, while permitting either reduced or stable levels of aboriginal and non-aboriginal resident hunting (Henheffer 2010; Moore 2010). In the Northwest Territories in Canada, where caribou outfitters have been facing ongoing declines in the hunting tags allocated to them since 2006, outfitters responded initially by varying the hunt experiences offered and by streamlining or closing parts of their businesses. Finally, some operators have launched a lawsuit against the government for its role in the management of caribou and of the caribou outfitting business (The Canadian Press 2011). In this case, the possibility of LCT is intricately linked with other uses, particularly the views and access interests of aboriginal and non-aboriginal residents, the effects of other stressors, the effects of natural processes on numbers, and the management decisions that attempt to take all of these aspects into account for current and future goals. Given the short time period in the case of caribou sport hunting, the journey, for the tourism sector, from the possibility of promoting LCT based on resource vulnerability to having no tourism opportunities at all can be quick, and one that occurs at a pace that might be beyond the ability of the industry to adapt.

## Disappearing indigeneity, emerging modernity: culture in last and first chance tourism

Nature and culture are intricately linked in the Arctic and tourism reflects this partnership. The frontier image of Arctic destinations promoted in much tourism is tempered by a growing recognition of the extent to which the Arctic is a homeland, too (Nutall, 2004). The Arctic is inhabited by numerous indigenous peoples, known also as aboriginal and first peoples. These populations and their communities are part of the modern world, but many of them also live a largely traditional life with values, interests and behaviour that can be quite distinct. All have a particular history and heritage that is the foundation for their culture and identity. Aboriginal communities are promoted in many parts of the Arctic as attractions available for tourist consumption; these communities represent an otherness that draws interest and the desire to experience a different culture. In one way, this approach itself may be viewed as LCT; some still live primarily in traditional ways, but are all engaged in processes of modernization or change. That is, at some point in the future, the traditions of yesterday that are showcased today will have turned into heritage and history.

This chapter now turns to an example of the interaction of culture and last chance/first chance tourism by examining a mid-1990s study of Sami tourism (Lyngnes and Viken 1997). In this study it was evident that some tourists thought the people they met in traditional clothing were acting a role in costume for the tourists, while other visitors wondered where the Sami communities were located, not realizing they were actually in the middle of these communities. The authors conclude that the ethnic signs of otherness in Sami communities are often hard to recognize or are hidden since the communities are modern and integrated (Lyngnes and Viken 1997). This suggests that some of the tourists may well have considered their visits as being too late to see the authentic Sami, while others may have been disappointed about the change to a modern life and landscape. However, it would be ethically inappropriate to label this kind of tourism experience as LCT. The Sami culture is strong, but as a strong culture it is also dynamic, like the rest of the Nordic cultures; the Sami are modern, changing, aboriginal peoples. This setting provides lots of first chance tourism experiences: new theatres, new folk rock music, new and rehabilitated traditions, new buildings and many new institutions integrating the Sami people in the Nordic welfare states.

Alongside growing tourism and other changes in society, cultural expressions change, but these are not due solely to climate change, but often reflect responses to the opportunities provided by tourism itself. Sometimes, the production of a tourism experience has replaced aspects of a traditional industry; for example, reindeer herding, harvesting or handicrafts. Such changes affect the culture: to keep reindeer for tourism is not the same as being a reindeer herder for subsistence. Also, with tourism, new cultural expressions emerge, often with no foundation in traditions or history. Yet, these changes may be seen as a function of a dynamic culture. From such a perspective, the disappearance and appearance of culture and its expressions is normal and therefore, once again, LCT is always tied to first time (new) possibilities.

Perhaps the most problematic side of LCT for these aboriginal cultures is the rhetoric it represents and its origin. It is not a term that comes from aboriginal cultures themselves and does not have inherent meaning or value to the culture. Rather this term comes from the dominant cultures outside, and can be seen when used to describe cultural tourism as a way of ‘othering’. It is claimed that such naming, as Laclau (2000) calls it, is not arbitrary, but reflects a power position. The tourism industry has been accused of colonizing aboriginal cultures, promoting and presenting the cultures for the sake of profit (Crick 1989; see Lemelin and Baikie this volume). In so doing, the cultures tend to be adapted, commercialized and commodified. Sometimes individuals who are aboriginal are themselves involved, and sometimes not. However, tourism can be a way of communicating the nature and dynamism of modern aboriginal life. In many Arctic places the aim is first chance, first class experiences provided by first peoples. Thus, there is strong normative element in the choice of terms.

## Concluding remarks

This chapter has focused on Arctic environments in a discussion of the emerging phenomenon of LCT. Due to the region's highly publicized vulnerability to changing environmental conditions and the rapid modernization of indigenous societies in the region, the Arctic is considered by many to epitomize the LCT concept. Examples of increased tourism interest are evident across the region and most often include increased cruise ship or small boat activity and enhanced Arctic wildlife viewing opportunities related to changing sea ice conditions. As suggested there are both risks and opportunities manifesting for local people, tourism operators, and tourists as the ideas surrounding LCT continue to be publicized by both media and academia.

Among many stakeholders, there is scepticism towards the LCT term even as there is interest in new ways of promoting tourism. To many, the term is simply rhetoric to promote tourism – another doomsday prophecy that people are trying capitalize upon. Similar to the Olsen *et al.* (this volume) study, researchers in the interview study described above were accused of setting this agenda and introducing the term simply by asking people about it. It is difficult to separate academic writing and knowledge seeking from economic or industrial actions and entrepreneurial responses to situations. Writing academically about a phenomenon can certainly be part of its creation. Thus, LCT should be considered a provocative term and one that requires careful contemplation by not only the research community, but also by the tourism industry, local communities and tourists themselves. Finally, LCT must be viewed within the context of tourism change and a dialectic that sees the opportunities provided by vulnerability as occurring alongside the opportunities enabled by changing conditions. The caveat for this last chance/first chance dialectic is the possibility that the sector or the individual businesses will be unable to adapt quickly enough to take advantage of either opportunity. Nonetheless, the vulnerability approach that is common to climate change research needs to be expanded to encompass an exploration of tourism responses potentially as behavioural or managerial adaptations and as innovations that exist alongside the dynamism of Arctic culture and nature.

## References

Aall, C. and Høyer, K.G. (2005) 'Tourism and climate change adaptation: the Norwegian Case', in C.M. Hall and J. Higham (ed.) *Tourism, recreation and climate change*, Clevedon, UK, and Buffalo, NY: Channel View Publications.

ACIA (Arctic Climate Impact Assessment) (2004) *Impacts of a warming Arctic*. Cambridge:

Cambridge University Press.

Brotton, J., and Wall, G. (1997) *The Bathurst Caribou Herd in a changing climate.* Climate Change Digest. Toronto: Government of Canada Atmospheric Environment Service.

Bone, R. (1992) *The geography of the Canadian North: Issues and challenges.* Toronto: Oxford University Press.

Buhasz, L. (2006) 'Northern underexposure', *Globe & Mail*, 1 July.

Chaturvedi, S. (1986) *The polar regions: A political geography.* Chichester, UK: Wiley.

Crick, M. (1989) 'Representations of international tourism in the social sciences: Sun, sex, sights, savings, and servility', *Annual Review of Anthropology* 18: 307–344.

Dawson, J., Lemelin, H., and Stewart, E.J. (2009) 'Last chance tourism: Risks and opportunities of an emerging market', *Conference Proceedings for the Travel and Tourism Research Association's (TTRA) Canada 2009 conference*, Guelph, Ontario, October 14–16.

Dawson, J., Maher, P., and Slocombe, S. (2007) 'Climate change, tourism and systems theory in the Canadian Arctic', *Tourism in Marine Environments* 4(2): 69–83.

Dawson, J., Stewart, E.J., Lemelin, H., and Scott, D. (2010) 'The carbon cost of polar bear viewing in Churchill, Canada', *Journal of Sustainable Tourism* 18: 319–336.

Eijgelaar, E., Thaper, C., and Peeters, P. (2010). 'Antarctic cruise tourism: the paradoxes of ambassadorship, "last chance tourism" and greenhouse gas emissions', *Journal of Sustainable Tourism* 18: 337–354.

Frew, E.A. (2008) 'Climate change and doom tourism: advertising destinations "before they disappear"', in J. Fountain and K. Moore, (eds.) *Recreating Tourism: Proceedings of the New Zealand Tourism & Hospitality Research Conference 2008*, 3–5 December, Hanmer Springs, New Zealand. CD Rom, Lincoln University, Canterbury, New Zealand.

Government of Northwest Territories (2006) *Caribou forever – Our heritage, our responsibility: A Barren-ground Caribou management strategy for the Northwest Territories 2006–2010.* Yellowknife: Northwest Territories Environment and Natural Resources.

Gunn, A. (1995) 'Responses of Arctic ungulates to climate change', in Peterson, D. and Johnson, D. (eds.) *Human ecology and climate change: People and resources in the far north,* Washington, DC: Taylor & Francis.

Hall, C.M., Müller, D., and Saarinen, J. (2009) *Nordic Tourism.* Bristol: Channelview Publications.

Henheffer, T. (2010) 'On a deadly trail', *Maclean's*, 29 November.

Hurtigruten (2010) Online: www.hurtigruten.no/Utils/ARCHIVE/Climate-Cruise-2010/Klimareisen-2010 (accessed 20 February 2011).

Jacobsen, J.K.S. (1994) *Arctic tourism and global tourism trends,* Research Report No. 37, Thunder Bay. Canada: Lakehead University Centre for Northern Studies.

Johnston, M.E. (2006) 'Impacts of global environmental change on tourism in the Polar Regions', in S. Gössling and C.M. Hall (eds.) *Tourism and Global Environmental Change*. London: Routlege.

Laclau, E. (2000) 'Power and social communication', *Ethical Perspectives* 7(2–3): 139–145.

Lemelin, R.H., and Johnston, M. (2008) 'Northern protected areas and parks', in P. Dearden, and R. Rollins (eds.) *Parks and protected areas in Canada: Planning and management*, 3rd edn. New York: Oxford University Press.

Lemelin, R.H., and Wiersma, E.C. (2007) 'Gazing upon Nanuk, the polar bear: the wildlife tourist gaze and ocular consumption in Churchill, Manitoba', *Polar Geography* 30(1–2): 37–53.

Lemelin, H., Dawson, J., Stewart, E.J., Maher, P., and Lück, M (2010) 'Last chance tourism: the doom, the gloom, and the boom of visiting vanishing destinations', *Current Issues in Tourism* 13(5): 477–493.

Lundmark, L. (2010) 'A holiday on ice on hold? Nature-based tourism and climate change in the Nordic north', in C.M. Hall and J. Saarinen (eds.) *Tourism and global change in Polar Regions: Climate environment and experiences*. New York: Routledge.

Lyngnes, S., and Viken, A. (1997) *Samisk Kultur og Turisme på Nordkalotten*. Oslo: BI.

Maher, P.T. (2010) 'Cruise tourist experiences and management implications for Auyuittuq, Sirmilik and Quttinirpaaq National Parks, Nunavut, Canada', in C.M. Hall and J. Saarinen (eds.) *Tourism and global change in Polar Regions: Climate environment and experiences*. New York: Routledge.

Maher, P.T., and Meade, D. (2008) *Cruise Tourism in Auyuittuq, Sirmilik and Quttinirpaaq National Parks*. Technical Report – ORTM Publication Series 2008–02. Prince George: UNBC ORTM Program.

Moore, O. (2010) 'Caribou crisis in Labrador', *Globe and Mail*, 23 December. Online: license.icopyright.net/user/view/FreeUse.act (accessed 13 January 2011).

Nesbitt, L. and Adamczewski, J. (2009) 'Decline and recovery of the Bathurst caribou herd: Workshops', 1/2 and 5/6 October, Yellowknife, *Summary Report*. Online: www.enr.gov.nt.ca/_live/document/content/Bathurst_Caribou_Modelling_Workshop_Report.pdf (accessed 13 January 2011).

NOAA (2006) 'State of the Arctic Report 2006'. Online: www.arctic.noaa.gov/soa2006/ (accessed 21 March 2011).

Nutall, M. (2004) 'Epilogue: Cultivating Arctic landscapes', in D G. Anderson and M. Nuttall (eds.) Cultivating Arctic landscapes: Knowing and managing animals in the circumpolar north. New York: Berghahn Books.

Øseth, E. (2010) Klimaendinger I Norsk Arktis. Konsekvenser for livet i nord. Report 136. Tromsø: Norwegian Polar Institute.

Ord C. (2010). 'Last chance tourism? Vanishing destinations and tourism's role in saving them'. Online: www.yourtravelchoice.org/2010/12/last-chance-tourism-vanishing-destinations-and-tourisms-roles-in-saving-them/ (accessed 15 March 2011).

Pagnan, J.L. (2003) 'The impact of climate change on Arctic tourism: A preliminary review', World Tourism Organization First International Conference on Climate Change and Tourism. Djerba, Tunisia, 9–11 April 2003. Online: www.world-tourism.org/sustainable/climate/pres/jeanne-pagnan.pdf (accessed 15 March 2011).

Saarinen, J., and Tervo, K. (2006) 'Perceptions and adaptation strategies of the tourism industry to climate change: the case of Finnish nature-based tourism entrepreneurs', International Journal of Innovation and Sustainable Development, 1(3): 214–228.

Saarinen, J., and Tervo, K. (2010) 'Sustainability and emerging awareness of a changing Climate', in C.M. Hall and J. Saarinen (eds.) Tourism and global change in Polar Regions: Climate environment and experiences. New York: Routledge.

Spitzer, A. (2007) 'The vanishing herds', *Up Here* 23(1): 42–46.

Stewart, E.J. (2009) *Resident attitudes toward tourism: Community-based cases from Arctic Canada,* unpublished thesis. University of Calgary, Alberta, Canada.

Stewart, E.J., and Draper, D.L. (2007) 'A collaborative approach to understanding local stakeholder perceptions of tourism in Churchill, Manitoba (Canada)', *Polar Geography* 30(1&2): 7–35.

Stewart, E.J., Howell, S.E.L., Draper, D., Yackel, J., and Tivy, A. (2007) 'Sea ice in Canada's Arctic: Implications for cruise tourism', *Arctic* 60(4): 370–380.

Stewart, E.J., Draper, D., and Dawson, J. (2011) 'Monitoring patterns of cruise tourism across Arctic Canada', in M. Lück, P. Maher, P. Stewart and E.J. Stewart (eds.) *Cruise tourism in the polar regions: Promoting environmental and social sustainability?* London: Earthscan Publications.

Sugden, D.E. (1989) 'The polar environment: Illusion and reality', *Ambio* 18(1): 2–5.

Tervo-Kankare, K., and Saarinen, J. (2011) 'Sub-Arctic environmental dimension. Climate change and adaptation strategies of the tourism industry in Northern Europe', in P. Maher, E.J. Stewart and M. Lück (eds.) *Polar tourism: Human, environmental and governance dimensions.* New York: Cognizant Communication.

The Canadian Press (2011) 'N.W.T. caribou hunters launch lawsuit'. Online: www.cbc.ca/news/canada/north/story/2011/03/14/nwt-caribou-outfitters-lawsuit.html (accessed 11 March 2011).

University of Calgary (2009) 'Climate change driving last chance tourism in the Arctic'. Online: www.ucalgary.ca/news/utoday/jan28–09/lastchance (accessed 19 March 2011).

Wang, N. (2000) *Tourism and modernity: A sociological analysis.* New York: Pergamon.

WWF (2008) 'Arctic climate impact science: an update since ACIA'. Oslo, Norway: International Arctic Program. Online: www.worldwildlife.org/climate/WWFBinaryitem8706.pdf (accessed 19 March 2011).

# 3 Last chance tourism in Antarctica

## Cruising for change?

*Machiel Lamers, Eke Eijgelaar, and Bas Amelung*

### Introduction

Over the last two decades tourism has grown rapidly in Antarctica. Tourist numbers have increased from a few thousand in 1990 to more than 45,000 in 2008, and recently declined again to less than 37,000 due to the global economic recession (IAATO 2010). Antarctic tourist activities are highly concentrated; spatially mainly in the Antarctic Peninsula region (see Figure 3.1) and seasonally between November and March. In addition, the product portfolio has diversified in various

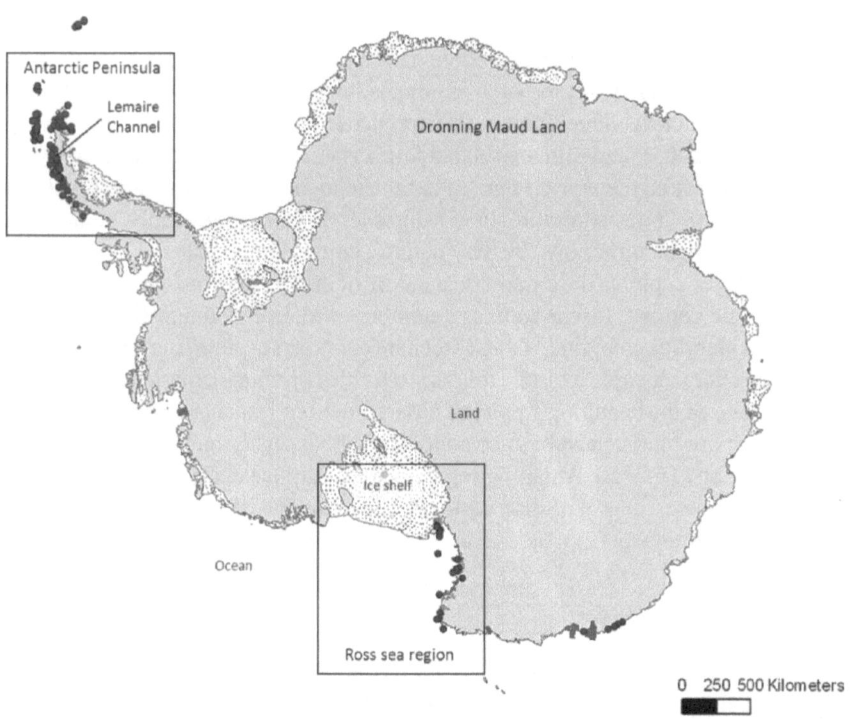


*Figure 3.1* Sites visited by tourists in Antarctica in the 2005/06 season

transport modes (e.g. aircraft, yachts, large cruise liners) and activities (e.g. kayaking, scuba diving, mountain climbing, marathon running).

Images of dramatic Antarctic landscapes and iconic wildlife have increasingly featured in media reports on global climate change, as well as Antarctic research activities crucial for understanding the global climate system, such as in Al Gore's award winning documentary 'An Inconvenient Truth'. Destinations affected by climate change are ever more seen as attractions for tourist visitation and the regions of the world that appear to be most affected are no exception (Eijgelaar *et al.* 2010; Lemelin *et al.* 2010). Numerous websites of news agencies, tour operators, tourism agencies, and others, urge potential tourists to go and see Antarctica before it changes or disappears (Abercrombie and Kent 2011; Geographical 2009; Polar Cruises 2011; Tickell 2007; Wiley 2008), to get a hands-on experience of climate change in Antarctica (Abercrombie and Kent 2011), to visit Antarctica as one of those places you must see before you die, or before the rest of the world gets there. The following citation is typical for the paradoxical nature of marketing Antarctica as a last chance destination:

> The best approach in terms of your own travel to Antarctica, might be if in doubt, go now before it changes further . . . and let's all hope changes here will slow, cease or be a long time coming.
>
> (IExplore 2011)

The citation illustrates that increased visitation to Antarctica is both a result and a driver of change in Antarctica, such as congestion (Haase *et al.* 2009; Lynch *et al.* 2009), infrastructural development (Bastmeijer *et al.* 2008), biological invasions (Frenot *et al.* 2005), and climate related impacts (Lamers and Amelung 2010). In destinations around the world, tourism has the tendency to 'kill the goose that lays the golden eggs' (Glasson *et al.* 1995). In other words, it inevitably reduces the quality of its main attractions, be they natural, cultural or social.

This chapter explores the paradoxical nature of last chance tourism in the changing Antarctic context. It will address a number of interrelated questions. To what extent is Antarctica changing? Does last chance tourism occur in the Antarctic context? To what extent are Antarctic tourism activities promoted, marketed, or driven by the image of the changing Antarctic environment or last chance destination? To what extent are tourists aware of or concerned about this issue? Do visitors (want to) take action? How are Antarctic regulatory institutions dealing with this issue? The following sections will discuss these questions based on relevant academic literature and policy papers, as well as a review of Antarctic visitor surveys.

## Changing Antarctica

Scientific sources report that the Antarctic region is subject to drastic changes occurring at multiple levels, from global climate change to local changes caused by increasing human activities (e.g. Vaughan *et al.* 2003; Tin *et al.* 2008). The Intergovernmental Panel on Climate Change (IPCC) has stated that with a global mean temperature increase of around 0.76°C since the end of the nineteenth century, the warming of the climate system is unequivocal (IPCC 2007). The IPCC also claims that the effects of climate change are most heavily felt at high latitudes. Antarctica is affected by climate change, but regional differences are profound. Over the past 50 years, mean annual temperatures in the Antarctic Peninsula have risen rapidly with almost 3°C (Vaughan *et al.* 2003; Turner *et al.* 2005), which makes this among the most rapidly warming regions in the world. The increasing temperatures have resulted in the loss of more than seven Antarctic ice shelves (Clarke and Harris 2003). In other instances, recent cooling trends have also been reported (Clarke and Harris 2003).

Climatic changes in Antarctica are likely to affect the development of tourism; for example, by decreased sea ice, longer summer seasons, and changes in wildlife populations (Lamers and Amelung 2010) (see Table 3.1 for an overview). The reduction of sea-ice extent is believed to increase tourism access to Antarctica, in particular by cruise vessels (Snyder 2007). New channels and islands have emerged from receding sea ice and glaciers, and has permitted the first circumnavigation of James Ross Island in the Antarctic Peninsula (Crosbie and Splettstoesser 1997). Devoid of spectacular scenery and wildlife, these new sites and areas are often deemed unattractive or unexciting for tourists. Currently the scenic Lemaire Channel in the Antarctic Peninsula region cannot be sailed at the start of the tourist season, but it may open up earlier in the season in the coming decades (see Figure 3.2). Direct impacts of climate change on the flora and fauna are difficult to predict since ecosystems are subject to multiple stressors and complex feedbacks. In addition, there is a growing threat of invasions by alien species due to shifting climatic zones and increasing human traffic for science and tourism, which increases the opportunity for introduction of alien species. Furthermore, slow reproduction rates may limit the possible relocation of native species. Such invasions have already occurred on many sub-Antarctic islands with detrimental consequences for native species (Frenot *et al.* 2005). Variations and fluctuations in sea-ice may affect the growth of krill and copepods, which is the foundation of the food chain and affects species, such as seals, penguins and whales. For example, populations of various penguin species have shifted considerably in the Antarctic Peninsula over the last decade (Lynch *et al.* 2008); shifts that have been related to climate change (Forcada *et al.* 2006).

*Table 3.1* General overview of Antarctic climate changes relevant for tourism

- The Antarctic Peninsula is warming rapidly and accelerated changes are projected
- Antarctic warming will have worldwide implications with regard to sea-level rise
- Vegetation zones are very likely to shift, causing wide-ranging implications
- Animal species' diversity, ranges, and distribution will change
- Increased precipitation and facilities face increasing exposure to storms
- Reduced sea ice is likely to increase opportunity for marine transport and access to resources
- Thawing ground will disrupt transportation, buildings and other infrastructure
- Elevated ultraviolet radiation will affect people, plants, and animals
- Multiple influences interact to cause impacts to people and ecosystems

*Figure 3.2* The Lemaire Channel at the end of the tourist season, March 2009

At the same time, tourism, and cruise ship tourism in particular, has grown rapidly in Antarctica. The effects of climate change and the growing role of tourism in Antarctica suggests a relation between the two sources of change and their combined effects (Stonehouse and Snyder 2007). Human activities in Antarctica affect the natural and human environment in which they are taking place (Tin *et al.* 2008), but are themselves also strongly influenced by this same environment. The impact of crowding in tourism is no exception (Lamers *et al.* 2008). Crowding can be a problem in any tourist destination, but in Antarctica even at low levels of visitation tourists can experience a strong sense of crowding, severely derogating from the area's key attractions. The presence of other tourists, ships, or infrastructure results in environmental impacts and makes the area less pristine (e.g. Bastmeijer *et al.* 2008), disturbs the wildlife (e.g. De Villiers 2008), and spoils the landscapes (e.g. Haase *et al.* 2009). Avoiding these effects by excluding human users is very difficult, if not impossible, particularly in a global commons such as Antarctica. Apart from impacts on tourism itself, Antarctic ecosystems as well as the research activities of the national programs may be affected by increased visitation. So far, no major biophysical impacts of tourism in the Antarctic have been determined (COMNAP 2003; Hofman and Jatko 2000; Stonehouse and Crosbie 1995), but monitoring programs are far from comprehensive (Enzenbacher 2007), and there are concerns about potentially cumulating impacts (Bastmeijer and Roura 2004). Also, the environmental impacts of catastrophic incidents, such as the 2007 sinking of the *MS Explorer* in the Antarctic Peninsula, remains largely unknown (Stewart and Draper 2008). As climate change and its direct and indirect effects occur, it is becoming increasingly difficult to separate impacts caused by tourism from other impacts.

Ironically, a consequence of increased visitation of Antarctica to witness climate change impacts is a further increase in greenhouse gas emissions that contribute to climate change. Antarctic cruise trips produce far higher amounts of $CO_2$ emissions per passenger than international trips on average. Apart from the cruise itself, return flights from home countries, generally in the Northern Hemisphere, contribute to these emissions (Amelung and Lamers 2007; Eijgelaar *et al.* 2010). Tourism development therefore depends on the acceptability of impacts at the local, regional and global scales. Research has already shown that the European consumer associates Arctic destinations with climate change (Huebner 2009), and a number of polar cruise operators suggest that the attention on climate change in the media is beneficial for their business (Lamers and Amelung 2010), but to what extent is last chance tourism in Antarctica occurring? And, does the net effect of images of calving Antarctic glaciers lead to more visitation? And could it also generate support for protective measures? The following sections address these two questions.

## Last chance tourism in Antarctica

Tourism academics have recently introduced and discussed the term 'last chance tourism', largely in a polar tourism context (Lemelin *et al.* 2010; Dawson *et al.* 2010; Eijgelaar *et al.* 2010; Hall and Saarinen 2010). Hence, the term's definition of Lemelin *et al.* (2010: 478) is very much applicable to Antarctica: 'a niche tourism market where tourists explicitly seek vanishing landscapes or seascapes, and/or disappearing natural and/or social heritage'. Antarctica can be seen as the last chance tourism destination par excellence. Being advertised and qualified frequently as the ultimate last frontier, Antarctica appears to be one of the only truly natural places left on earth (e.g. Bastmeijer *et al.* 2008). But even Antarctica's frontier image may be vanishing. Stonehouse and Crosbie (1995) questioned the appropriateness of marketing Antarctic cruises as 'expeditions' as early as the mid-1990s, when cruise ships were already having difficulty keeping out of each other's sight in order to maintain an unspoilt seascape image for their passengers.

With 36,881 visitors in the 2009/10 season (IAATO 2010), Antarctic tourism can still be considered a niche market compared to other destinations. In fact, the economic recession currently appears to have reversed the rapid growth that has characterized the Antarctic tourism market since the early 1990s. The International Maritime Organization's (IMO) ban on the use and carriage of heavy fuel oil in Antarctic waters is expected to lead to a further fall in visitor numbers, due to fewer visits of large cruise-only ships (see Figure 3.3). For last chance tourists who wish to see Antarctica before it is irrevocably altered or spoiled by the presence of other cruise ships the declining number of IAATO member voyages – from 308 in 2007/08 to 239 in 2009/10 (IAATO 2010) – would seem to be good news. The IMO ban has even spurred a temporary kind of last chance marketing, as some companies have been advertising that the 2010/11 season might be the last

*Figure 3.3* Large cruise ship at King George Island, February 2008

chance to visit Antarctica on a luxury cruise ship (Cruise Holidays 2010; Fun For Less Tours 2010). The last tourist season of the ice breaker *Kapitan Khlebnikov*, which will return to escort duties in the Russian Arctic, is also marked by a series of special last chance trips to the Arctic and Antarctic (Quark Expeditions 2009).

However, it is not unthinkable that a number of potential large ship tourists will switch to smaller expedition vessels and cause an increase in the number of landings. IAATO projections for the coming seasons seem to acknowledge this, as from 2010/11 onwards, seaborne tourism with landings in Antarctica is expected to increase rapidly again (IAATO 2010). Thus the Antarctic landscape that tourists see during landings could still be at risk at frequently visited sites. Several historic sites on the Antarctic Peninsula and the South Shetland Islands have, and will likely continue to receive, over 10,000 visitors a year each, making them 'mass tourism destinations' by Antarctic standards (Roura 2010; Serrano 2007). These mass visits are concentrated in a short season, thereby increasing temporal pressures on Antarctic ecosystems (Hall and Wouters 1995). As one concerned editor of an online cruise guide noted in his 'Best Cruise Trends for 2008':

> Another trend for 2008 is cruises to Antarctica. I am hesitant to promote these cruises as the area is getting overcrowded, and some of the ships are too large to be going there. But the way things are going, the years ahead could be your last chance to see Antarctica in any kind of pristine condition. Who knows what may happen in the future – over-commercialization, for example, or a world agreement to shut it down to tourism completely to save it from overexposure.
>
> (Motter 2008)

From the above, tourism impacts on the Antarctic landscape and seascape appear to be more important factors in changing Antarctica, and driving last chance tourism, than other possible causes, such as climate change. However, climate change could play an important role in driving the development of tourism. For instance, large cruise multinationals, such as Carnival Corporation, see a potential demand for visiting remote areas impacted by climate change, such as Alaska and Antarctica (Carbon Disclosure Project (CDP) 2008). This increased demand may have partly been caused by travel journalists and other online media, producing must-see lists or books of endangered remote places that frequently include Antarctica (Hughes 2008; JWT 2007; Lisagor and Hansen 2008; Ruiz 2008; Shipman 2007; Smith 2008, Struck 2004). In these articles the rapid growth of Antarctic tourism is often attributed to its 'endangered status' (see also Hall and Saarinen 2010). As an online travel journalist states:

> Travelers seek out imperiled destinations and try to experience their grandeur before they vanish. The trend is best embodied by a surge in the number of travelers to the Galápagos Islands and Antarctica, both besieged by changes to their ecosystems.
>
> (Ruiz 2008)

Nuttall (2010), a historian onboard the *Explorer* during the 2006/07 season, remarks that many Antarctic tourism companies 'play on the increased awareness of global environmental issues supposedly expressed by passengers as one of the main reasons for travelling to Antarctica' (Nuttall 2010: 206). He goes on to explain that, besides the usual information on the continent, passengers are filled in about Antarctica's place within global environmental issues. Visitors 'are made to feel that they are travelling in an environment that is itself at risk. . . . under threat of disappearing, as the ice melts and yet another ice shelf collapses' (Nuttall 2010: 210). The assumed relationship between the dramatic reports on the state of the Antarctic environment and traveler's motivations for visiting the continent is explored in more detail in the following section.

## Visitor awareness and action

To investigate the presumed link between the changes in Antarctica and the motivation of tourists to visit Antarctica, key issues are the awareness of Antarctic tourists of these changes, the extent to which awareness provides a motive for travel, and possibly for a change in environmental behavior. We shed some light on these questions through an analysis of Antarctic visitor surveys that were conducted among Antarctic passengers between 1992 and 2009 (see Table 3.2 for basic features of selected visitor surveys). Surveys were held before or after visiting the Antarctic region, or both. They were conducted on cruise passengers of different origin, their destination usually being the Antarctic Peninsula. Despite varying countries of origin, the visitor profile of most of these surveys is very similar: maturely aged, highly educated, and retired or professional. However, we have to

*Table 3.2* Basic features of selected visitor surveys

| *Source* | *Time* | *Mode* | *Sample* | *Origin* | *Destination* | *Season* |
|---|---|---|---|---|---|---|
| Cessford and Dingwall (1996) | Pre/post | Ship | 458 | Mix | NZ sub-Ant. Isl. | 92/93, 93/94 |
| Bauer (2001) | Pre/post | Ship | 159/138 | Mix | Antarctica | 94/95, 95/96 |
| | | Flight | 484 | AUS | | 94/95, 95/96 |
| | | Flight | 1312 | | | 96/97 |
| Bauer and Dowling (2006) | – | – | 55 | – | Antarctica | 00/01 |
| Maher *et al.* (2006) | Pre | Ship | 55 | 2/3 NZ | Ross Sea | 02 |
| Powell *et al.* (2008) | Pre/post | Ship | 288/266/ 125 | 60% US | Peninsula | 02/03, 03/04 |
| Maher (2007) | Mix | Ship | – | Mix | Ross Sea | 02/03, 03/04 |
| Vereda (2010) | Pre/post | Ship | 141/141 | Mix | Peninsula | – |
| Maher *et al.* (2010) | Post | Ship | 38 | 2/3 US | Peninsula | 07/08 |
| Eijgelaar *et al.*(2010) | Post | Ship | 151 | 98% D/CH/A | Peninsula | 08/09 |

be careful with comparing results from differently designed and conducted surveys, as the explanatory power of such an analysis is limited. The results described in the following paragraphs are a means to start to explore visitor awareness and action in relation to the changing Antarctic.

### *Motivations*

In most surveys ship-borne visitors consider wildlife, scenery, adventure and remoteness (sometimes phrased differently) to be the most important motives for traveling to Antarctica (Bauer 2001; Cessford and Dingwall 1996; Maher *et al.* 2010; Maher *et al.* 2006; Powell *et al.* 2008; Vereda 2010). The two most recent studies have explicitly included the 'last chance' motivation in their surveys (Eijgelaar *et al.* 2010; Maher *et al.* 2010). Maher *et al.* asked for a ranking (on a 1–5 scale; strong to weak) of 21 motivational categories, whereas Eijgelaar *et al.* allowed for multiple answers (eight motives plus one open). Both the question type and the number of available motives could explain why 'see the Antarctic before it is gone' ended up as being more important (fourth ranking motive; 33 per cent) with Eijgelaar *et al.*'s respondents, compared with Maher *et al.*'s respondents, who ranked 'last chance' as a rather unimportant motive for their trip (mean score 3.58). Interestingly enough, Maher *et al.*'s respondents actually appeared to be more aware of climate change impacts in Antarctica and tourism impacts on climate change than Eijgelaar *et al.*'s respondents (see below).

### *Awareness of impact*

Antarctic visitors often show high levels of general environmental concern (Maher 2010). It might therefore be conjectured that they are also sensitive to environmental quality and change. The tourists' level of sensitivity is difficult to establish, however, because no explicit questions on perceived environmental impacts and change were included in surveys until recently. These recent surveys suggest that climate change is becoming an important issue for visitors and tourism professionals in Antarctica. For example, 3 per cent of Vereda's (2010) respondents expected to gain awareness about climate change in the Antarctic. Almost half of the respondents of Eijgelaar *et al.* (2010) claimed to have actually experienced the effects of climate change during their cruise. In the Maher *et al.* (2010) survey, 80 per cent agreed on the statement that 'climate change is going to impact Antarctica'. So far, climate change and other forms of environmental change and impacts do not appear to have a noticeable impact on tourist satisfaction. For example, 93 per cent of the Bauer and Dowling (2006) sample agreed to the statement that the condition of the Antarctic was excellent.

### *Awareness of personal contributions to impact*

Cessford and Dingwall (1996) found that on the sub-Antarctic Islands (Southern Ocean islands north of 60° south latitude) where landings were made, around 30 per cent acknowledged that their visit made an impact (i.e. mainly trampling). For Bauer's (2001) ship-borne and overflight visitors, tourism impacts were not an issue. Respondents of Bauer and Dowling (2006) commented that humans 'did not have the right to alter' the Antarctic, yet at the same time they acknowledged that the only way to truly appreciate the continent was by seeing it first-hand. This would imply that they believed their visits would not alter the Antarctic in any way. Maher (2007) reports that before the trip, 72 per cent of the respondents believed their visit would both harm and support Antarctica. The majority of the respondents believed that any negative impact would be outweighed by their own support for conservation of the continent. After the visit, the number of those who shared this belief grew further. Only 6 per cent of the respondents thought that their visit had only negative impacts. It is generally believed that a first-hand experience leads to growing awareness and support for Antarctic conservation which in turn outweighs the impacts. This view is actually promoted by IAATO through the term 'ambassadorship' (e.g. IAATO 2009; Maher *et al.* 2003); see also Lemieux and Eagles this volume). Eighty per cent of Maher's (2007) respondents felt like ambassadors themselves, as opposed to only 1.4 per cent of Bauer's (2001) post-visit respondents. This discrepancy could be explained by differences in the location of the samples (i.e. Ross Sea region vs. Antarctic Peninsula) and in the sample population. Whereas Bauer's sample population exclusively consisted of tourists, Maher's sample also included visitors of the New Zealand Antarctic program and students on an educational expedition.

Recent surveys have also addressed aspects related to climate change. In Maher's *et al.* (2010) survey, two-thirds of the respondents agreed on the human contribution to climate change, with 61 per cent pointing at air travel as a contributor, and 54 per cent at cruise travel. To some extent, Eijgelaar's *et al.* (2010) survey showed different results. While nearly all respondents believed that climate change existed, only 42 per cent connected this to anthropogenic causes. Here, 59 per cent did not believe that traveling would have a significant impact on climate change, 21 per cent believed the contrary, and 20 per cent did not know. Sixty per cent also believed that the emissions caused by their own Antarctic cruise trip would not differ much from the average international holiday trip.

## Changes in attitudes and behavior

From the above, visitors appear to be only marginally aware of their own direct impact on the Antarctic environment and of their indirect impact through the greenhouse gas emissions caused by their trips. They also do not appear too concerned about it, so that a significant and voluntary change toward more environmentally friendly behavior is unlikely to materialize in the near term. Indeed, Maher (2007) found that most of his respondents considered themselves ambassadors, but did not intend to change their future travel behavior. Pro-environmental intentions and behaviors only appeared to increase directly after a visit, dissipating again after a few months (Powell *et al.* 2008). Many Antarctic tourists also reported plans to visit a number of other ecotourism destinations (Maher 2010; Eijgelaar *et al.* 2010; Maher *et al.* 2010; Bauer 2001). Thus it is also not likely that they intend to reduce, or are aware of, their own travel emissions. In addition to spontaneous attitudinal and behavioral changes of Antarctic tourists, a range of management and policy frameworks apply to Antarctic tourism, which will be discussed in the following section.

## Change-oriented management and regulation

Antarctica is a tourism destination with a global appeal and is governed by multiple actors, with regard to image building, marketing (e.g. tourism industry), and regulating impacts (e.g. state actors, NGOs, tourism industry). Regulatory action taken to tackle impacts and changes caused by or affecting tourism in Antarctica is limited, which is mainly due to the unique Antarctic governance context (Bastmeijer and Roura 2004; Molenaar 2005). The level of control over Antarctic tourism that the Antarctic Treaty System (the consortium of states that jointly manage Antarctic human activities and natural resources) currently exercises is considered low. A range of regulatory limitations (including delays in implementation) are identified in the literature discussing Antarctic governance in general, and tourism regulation and management in particular (Bastmeijer and Roura 2004; Hemmings and Roura 2003; Kriwoken and Rootes 2000).

In many states, the authorization of individual tourism trips or single companies' activities is informed by pre-seasonal Environmental Impact Assessments (EIA), as laid out in the Madrid Protocol, the main international legislation regulating tourism. However, the actual, large-scale, and longer-term effects of tourism, as well as ethical issues, tend to be ignored in these EIAs (Bastmeijer and Roura 2004; Hemmings and Kriwoken 2010; Hemmings and Roura 2003; Kriwoken and Rootes 2000). For example, environmental audits and multiple-year assessments are not performed, and global greenhouse gas emissions are not included. Onsite monitoring and enforcement of regulatory mechanisms for Antarctic tourism is extremely difficult because of the remoteness and vastness of the area that would have to be covered (Molenaar 2005; Tracey 2001). Nevertheless, some regulatory action has been taken for improving information delivery and mitigating some of the negative effects of tourism, such as the establishment of codes of conduct, pre- and post-trip notification, information exchange between Antarctic Treaty Parties, compulsory insurance and contingency planning, and site-specific guidelines.

Currently, much of the impetus towards environmentally sound tourism regulation comes from the International Association of Antarctica Tour Operators (IAATO), an industry association that unites most, but not all, commercial Antarctic tour operators and coordinates their activities. IAATO actively tackles several of the tourism-related changes and impacts, with congestion being a good example. By scheduling site visits before the season starts, one ship visits one site at one time, and visitation is spread across the available sites. At particularly sensitive sites, visits are controlled by means of site-guidelines. Some National Antarctic Programs (NAP) have set a quota for the number of ships that are annually allowed to visit their research stations, such as the US Palmer Station, and local visitation rules are enforced at individual stations to manage the flow of tourists. Standards for operating landings have been developed by IAATO, which state that ships carrying more than 500 passengers are not allowed to land people onshore, and that no more than 100 people are allowed to be onshore at any one time, with at least one guide per 20 tourists. These industry standards were recently adopted by the ATS as a non-binding resolution (ATCM 2007). Whether these operational rules and site-specific guidelines (and other spatial management instruments, such as Antarctic Specially Managed Areas) are climate and future proof remains to be seen and requires further research.

Potential and actual environmental impacts of Antarctic tourism have long been recognized in the literature and IAATO has been exemplary in their adoption of preventive measures for landing operations, such as the introduction of invasive species. Compulsory informative lectures are given prior to onshore and offshore activities in Antarctic territory to increase tourist awareness of littering, souveniring (taking stones, bones, artifacts, etc.), disturbing wildlife, trampling flora, damaging historic constructions, and the importance of clean equipment. Guides and expedition staff, as well as fellow-tourists, play a crucial role in enforcing these rules. Further, boot-washing practices with specially selected detergents are compulsory at entering the zodiacs and returning to the ship after an onshore visit.

Scholarly progress in the determination of global impacts of Antarctic tourism has been made (Amelung and Lamers 2007; Eijgelaar *et al.* 2010), but the Antarctic policy response has been limited. Although it is true that cutting back on greenhouse gas emissions is a global policy process, policy actions are needed on all levels to determine and tackle global impacts. It has been argued that the lack of reliable emission data calls for the inclusion of greenhouse gas inventories in the EIAs required by the Madrid Protocol (Amelung and Lamers 2007). No major actions have been noted by the Antarctic Treaty Parties in this direction.

Apart from data voluntarily supplied by tour operators (Eijgelaar *et al.* 2010) and despite its appearance in a policy paper (Norway and United Kingdom 2008), the requirement of greenhouse gas inventories or other types of reporting mechanisms is not recommended in the most recent reports. Recently, however, the Co-Chair of the Antarctic Treaty Meeting of Experts emphasized the importance of reducing carbon footprints and fuel consumption related to Antarctic activities. IAATO is taking up the gauntlet for tourism through its Climate Change Working Group, trying to reduce the carbon footprints of their ships (ATME 2010). Still, Antarctic tourism stakeholders appear to be more concerned about tourism being affected by climate change than about tourism's own contribution to climate change (Liggett *et al.* 2011). Stakeholder concern about tourism impacts does seem to have increased when compared to responses in Bauer's (2001) study, but is mainly directed toward impacts from the possible development of land-based facilities, accidents and an overall increase in scale.

Further, monitoring programs have been developed by NGOs, such as Oceanites (Naveen *et al.* 2000), and some of the NAPs (Australia 2004). However, to date, monitoring is performed on a voluntary basis and lacks a comprehensive and consistent approach (Australia 2004). We argue that the development of a more structured and programmatic approach to scientifically monitoring changes and human impacts, including greenhouse gas emissions, provide an important step in raising awareness of Antarctic Treaty Parties, the tourism industry, and tourists, of the impacts of tourism in a changing Antarctic environment.

## Conclusion

In the last two decades, tourism to Antarctica has grown spectacularly, and global environmental change has become a major societal issue around the globe. This chapter has explored the hypothesis that these two developments may be related. Images of calving Antarctic glaciers and other phenomena that are attributed to global change may have incited people to visit Antarctica 'before it is too late', engaging in last chance tourism. The chapter approaches the issue by asking whether Antarctica is undergoing major and irreversible change, whether (perceived) change has led to the development of last chance products for Antarctic tourism, whether tourists are aware of their contribution to the changing Antarctic, and whether policy initiatives are being taken to address this phenomenon.

A general review of media publications and advertisements on corporate websites suggests that both global and local changes are encouraging the appearance of last chance tourism products, with the last chance discourse being centered around the disappearance of ecosystems (visit before their grandeur is gone) and congestion (visit before it is too crowded).

Even though concern about environmental change may be a motive for visiting Antarctica, an analysis of recent visitor surveys demonstrates that the actual awareness among tourists of the occurring changes is limited. The same holds true for the awareness of their own roles as Antarctic tourists in causing local and global change. Tourists are not very conscious of the contribution of their own travel-related emissions to climate change, and do not reflect on options to reduce the environmental impact of their trip, let alone on the ethics of their visit (see also Dawson *et al.* 2010). On the contrary, visitors are generally very satisfied about their trip, would do it again and/or recommend it to others.

The tourism industry is taking the lead in developing site-specific guidelines to minimize and monitor impacts and diffuse visitation, but the more fundamental and paradoxical issues such as the marketing and advertisement of last chance tourism opportunities are not addressed. Tourism has often been accused of killing the goose that lays the golden egg, and (last chance) tourism to Antarctica appears to be a prime example of this. To stick with the metaphor: tourism not only disturbs the goose directly, but also affects the goose's habitat indirectly through its contribution to global warming. Dealing with this paradox is not an easy task, but making tourists aware of the environmental consequences of their own travel patterns may be a meaningful first step. The carbon-labeling of travel packages, including those to Antarctica, would help in making this visible and could lead to more environmentally conscious holiday choices (Gössling 2011).

## References


Abercrombie and Kent (2011) 'Fighting Climate Change in Antarctica' Online: www.abercrombiekent.com/news/press_releases/Fighting-Climate-Change-in-Antarctica.cfm?&bcid=1575> (accessed 1 October 2011).

Amelung, B. and Lamers, M. (2007) 'Estimating the greenhouse gas emissions from Antarctic tourism', *Tourism in Marine Environments* 4: 121–34.

ATCM (2007) 'Final Report – Part II: Measures, Decisions and Resolutions', Antarctic Treaty Consultative Meeting XXX, New Delhi, India.

ATME (2010) 'Co-Chairs' Report from Antarctic Treaty Meeting of Experts on Implications of Climate Change for Antarctic Management and Governance', in ATME (ed.) *Antarctic Treaty Meeting of Experts*. Svolvær, Norway.

Australia (2004) 'ATME#16 An approach to monitoring for non-government activities in Antarctica', Antarctic Treaty Meeting of Experts on Tourism and Non-governmental Activities, Tromso, Norway.

Bastmeijer, C. and Roura, R. (2004) 'Regulating Antarctic Tourism and the Precautionary Principle', *The American Journal of International Law* 98: 763–81.

Bastmeijer, C., Lamers, M. and Harcha, J. (2008) 'Permanent Land-based Facilities for Tourism in Antarctica: The need for regulation', *Review of European Community and International Environmental Law (RECIEL)* 17: 84–99.

Bauer, T. G. (2001) *Tourism in the Antarctic: opportunities, constraints and future prospects,* New York: Haworth Hospitality Press.

Bauer, T. and Dowling, R. K. (2006) 'The Antarctic cruise industry', in R. K. Dowling (ed.) *Cruise ship tourism,* Wallingford, UK: CABI.

Carbon Disclosure Project (CDP) (2008) 'Respondent CDP 2008: Carnival Corp.', Carbon Disclosure Project.

Cessford, G. R. and Dingwall, P. R. (1996) 'Tourist visitors and their experiences at New Zealand subantarctic islands', *Science & Research Series*, Wellington, New Zealand: Department of Conservation.

Clarke, A. and Harris, C. (2003) 'Polar marine ecosystems: major threats and future change', *Environmental Conservation* 30: 1–25.

COMNAP (2003) 'IP37 Interaction Between National Operators, Tourists and Tourism Operators', Antarctic Treaty Consultative Meeting XXVI, Madrid, Spain.

Crosbie, K. and Splettstoesser, J. (1997) 'Circumnavigation of James Ross Island, Antarctica', *Polar Record* 33: 341.

Cruise Holidays (2010) 'Last Chance for Large-Ship Cruises to Antarctica?' Online: www.elitecruisevacations.com/resource-center/cruise-related-articles/destinations/antarctica-last-chance-for-large-ships.aspx (accessed 15 December 2010).

Dawson, J., Stewart, E. J., Lemelin, R. H. and Scott, D. (2010) 'The carbon cost of polar bear viewing tourism in Churchill, Canada', *Journal of Sustainable Tourism* 18: 319–36.

De Villiers, M. (2008) 'WP12 Review of recent research into the effects of human disturbance on wildlife in the Antarctic and sub-Antarctic region', Antarctic Treaty Consultative Meeting XXXI, Kyiv, Ukraine.

Eijgelaar, E., Thaper, C. and Peeters, P. (2010) 'Antarctic cruise tourism: the paradoxes of ambassadorship, "last chance tourism" and GHG emissions', *Journal of Sustainable Tourism* 18: 337–54.

Enzenbacher, D. (2007) 'Antarctic Tourism Policy-making: Current Challenges and Future Prospects', in G. Triggs and A. Riddell (eds.) *Antarctica: legal and environmental challenges for the future,* London: The British Institute of International and Comparative Law.

Forcada, J., Trathan, P., Reid, K., Murphy, E. J. and Croxall, J. P. (2006) 'Contrasting population changes in sympatric penguin species in association with climate warming', *Global Change Biology* 12: 411–23.

Frenot, Y., Chown, S., Whinam, J., Selkirk, P., Convey, P., Skonicki, M. and Bergstrom, D. (2005) 'Biological invasions in the Antarctic: extent, impacts and implications', *Biological Review* 80: 45–72.

Fun For Less Tours (2010) 'South America & Antarctica Cruise'. Online: funforlesstours.com/tour/89/ (accessed 17 November 2010).

Geographical (2009) 'Here today, gone tomorrow (get the t-shirt)'. Online: www.geographical.co.uk/Magazine/Polar_tourism_-_Nov_09.html (accessed 1 October 2010).

Glasson, J., Godfrey, K. and Goodey, B. (1995) *Towards visitor impact management,* Aldershot: Ashgate Publishing.

Gössling, S. (2011) *Carbon management in tourism: mitigating the impacts on climate change,* Abingdon, UK: Routledge.

Haase, D., Lamers, M. and Amelung, B. (2009) 'Heading into uncharted territory? Exploring the institutional robustness of self-regulation in the Antarctic tourism sector', *Journal of Sustainable Tourism* 17: 411–30.

Hall, C. M. and Saarinen, J. (2010) 'Last chance to see? Future issues for polar tourism and change', in C. M. Hall and J. Saarinen (eds.) *Tourism and change in polar regions: climate, environment and experience,* Abingdon, UK: Routledge.

Hall, C. M. and Wouters, M. (1995) 'Issues in Antarctic tourism', in C. M. Hall and M. E. Johnston (eds.) *Polar tourism: tourism in the Arctic and Antarctic regions,* Chichester, UK: John Wiley & Sons.

Hemmings, A. and Roura, R. (2003) 'A square peg in a round hole: fitting impact assessment under the Antarctic Environmental Protocol to Antarctic tourism', *Impact Assessment and Project Appraisal* 21: 13–24.

Hemmings, A. D. and Kriwoken, L. K. (2010) 'High level Antarctic EIA under the Madrid

Protocol: state practice and the effectiveness of the Comprehensive Environmental Evaluation process', *International Environmental Agreements* 10: 187–208.

Hofman, R. and Jatko, J. (2000) 'Assessment of the Possible Cumulative Environmental Impacts of Commercial Ship-Based Tourism in the Antarctic Peninsula Area', Workshop, La Jolla, California, 7–9 June 2000.

Huebner, A. (2009) 'Tourist images of Greenland and the Arctic: a perception analysis', *Polar Record* 45: 153–66.

Hughes, H. (2008) *Frommer's 500 places to see before they disappear,* Hoboken: Wiley.

IAATO (2009) 'About IAATO: Bylaws'. Online: www.iaato.org/bylaws.html (accessed 30 September 2009).

IAATO (2010) 'IP113 IAATO Overview of Antarctic Tourism: 2009–10 season and preliminary estimates for 2010–11 and beyond', Antarctic Treaty Consultative Meeting XXXIII, Punta del Este, Urugay.

IExplore (2011) 'Are the Glaciers Melting? Understanding the Impact of Climate Change on Antarctica'. Online: www.iexplore.co.uk/experiences/antarctica/articles/are-the-glaciers-melting? (accessed 10 January 2011).

IPCC (2007) 'Summary for Policy Makers and Technical Summary', in S. Solomon (ed.) *Climate Change 2007: The Physical Science Basis. Contribution of Working Group 1 to the Fourth Assessment Report of the Intergovernmental Panel on Climate Change,* Cambridge: Cambridge University Press.

JWT (2007) 'Eighty things to watch in 2008'. Online: www.jwt.com/pdf/news/things-towatch2008.pdf (accessed 11 August 2009).

Kriwoken, L. K. and Rootes, D. (2000) 'Tourism on Ice: environmental impact assessment of Antarctic tourism', *Impact Assessment and Project Appraisal* 18: 138–50.

Lamers, M. and Amelung, B. (2010) 'Climate Change and its Implications for Cruise Tourism in the Polar Regions', in M. Lueck, P. Maher and E. Stewart (eds.) *Cruise Tourism in Polar Regions: Promoting Environmental and Social Sustainability?* London: Earthscan.

Lamers, M., Haase, D. and Amelung, B. (2008) 'Facing the Elements: Analysing trends in Antarctic Tourism', *Tourism Review* 63: 15–27.

Lemelin, H., Dawson, J., Stewart, E. J., Maher, P. and Lueck, M. (2010) 'Last-chance tourism: the boom, doom, and gloom of visiting vanishing destinations', *Current Issues in Tourism* 13: 477–93.

Liggett, D., McIntosh, A., Thompson, A., Gilbert, N. and Storey, B. (2011) 'From frozen continent to tourism hotspot? Five decades of Antarctic tourism development and management, and a glimpse into the future', *Tourism Management* 32: 357–66.

Lisagor, K. and Hansen, H. (2008) *Disappearing destinations: 37 places in peril and what can be done to help save them,* New York: Vintage Books.

Lynch, H., Crosbie, K., Fagan, W. and Naveen, R. (2009) 'Spatial patterns of tour ship traffic in the Antarctic Peninsula region', *Antarctic Science* 22: 123–30.

Lynch, H. J., Naveen, R. and Fagan, W. F. (2008) 'Censuses of penguin, blue-eyed shag and southern giant petrel populations on thee Antarctic Peninsula, 2001–2007', *Marine Ornithology* 36: 83–97.

Maher, P. T. (2007) 'Advocating for Antarctic wilderness: Short-term visits and human values', in A. Watson, J. Sproull and L. Dean (eds.) *Science and stewardship to protect and sustain wilderness values: Eighth World Wilderness Congress symposium. Proceedings RMRS-P-49,* Fort Collins, CO: US Dep of Agriculture, Forest Service, RMRS.

Maher, P. T. (2010) 'Footsteps on the ice: visitor experiences in the Ross Sea region, Antarctica', unpublished PhD thesis, Lincoln University.

Maher, P. T., Johnston, M. E., Dawson, J. P. and Noakes, J. (2010) 'Risk and a changing environment for Antarctic tourism', *Current Issues in Tourism*, 21 July (iFirst): 1–13.

Maher, P. T., McIntosh, A. J. and Steel, G. D. (2006) 'Examining Dimensions of Anticipation: Inputs Prior to Visiting the Ross Sea Region, Antarctica', *Tourism in Marine Environments* 2: 51–63.

Maher, P. T., Steel, G. and McIntosh, A. (2003) 'Antarctica: tourism, wilderness, and "ambassadorship"', in A. Watson and J. Sproull (eds.) *Science and stewardship to protect and sustain wilderness values: Seventh World Wilderness Congress symposium*, Port Elizabeth, South Africa: USDA Forest Service.

Molenaar, E. J. (2005) 'Sea-Borne Tourism in Antarctica: Avenues for Further Intergovernmental Regulation', *International Journal for Marine and Coastal Law* 20: 247–95.

Motter, P. (2008) 'Best Cruise Trends for 2008', CruiseMates. Online: www.cruisemates.com/articles/feature/bestcruisetrends2008–012808.cfm#axzz18C1jziSq (accessed 15 December 2010).

Naveen, R., Forrest, S., Dagit, R., Blight, L., Trivelpiece, W. and Trivelpiece, S. (2000) 'Censuses of penguin, blue-eyed shag, and southern giant petrel populations in the Antarctic Peninsula region, 1994–2000', *Polar Record* 36: 323–34.

Norway and United Kingdom (2008) 'WP35 Antarctic Climate Change Issues', Antarctic Treaty Consulative Meeting XXXI, Kyiv, Ukraine.

Nuttall, M. (2010) 'Narratives of history, environment and global change: expeditioner-tourists in Antarctica', in C. M. Hall and J. Saarinen (eds.) *Tourism and change in polar regions: climate, environment and experience,* Abingdon, UK: Routledge.

Polar Cruises (2011) 'Polar facts: Cruising Antarctica'. Online: www.polarcruises.com/antarctica/articles/cruising-antarctica_10/ (accessed 10 January 2011).

Powell, R. B., Kellert, S. R. and Ham, S. H. (2008) 'Antarctic tourists: ambassadors or consumers?', *Polar Record* 44: 233–41.

Quark Expeditions (2009) 'End of An Era Begins: Announcing Kapitan Khlebnikov's Retirement'. Online: www.quarkexpeditions.com/End-of-An-Era-Begins-Announcing-Kapitan-Khlebnikovs-Retirement (accessed 23 February 2011).

Roura, R. (2010) 'Cultural heritage tourism in Antarctica and Svalbard: patterns, impacts and policies', in C. M. Hall and J. Saarinen (eds.) *Tourism and change in polar regions: climate, environment and experience*, Abingdon, UK: Routledge.

Ruiz, R. (2008) 'Endangered natural wonders worth seeing', Forbes.com. Online: www.forbes.com/2008/09/15/travel-endangered-ecosystems-forbeslife-cx_rr_0915travel.html (accessed 16 December 2010).

Serrano, E. (2007) 'Tourism in the South Shetland Islands: recent changes in local destinations and activities', *Tourism in Marine Environments* 4: 221–35.

Shipman, T. (2007) 'Nature's "doom" is tourist boom', Telegraph.co.uk. Online: www.telegraph.co.uk/earth/earthnews/3319800/Natures-doom-is-tourist-boom.html(accessed 18 August 2009).

Smith, J. (2008) 'Endangered destinations: places, like species, can vanish forever. A look at some unique, imperiled treasures', *U.S. News & World Report,* May 26–June 2: 36–43.

Snyder, J. (2007) 'Tourism in the Polar Regions. The Sustainability Challenge', Paris: United Nations Environmental Programme/The International Society.

Stewart, E. and Draper, D. (2008) 'The Sinking of the MS Explorer: Implications for Cruise Tourism in Arctic Canada', *Arctic* 61: 224–8.

Stonehouse, B. and Crosbie, K. (1995) 'Tourist Impacts and Management in the Antarctic Peninsula Area', in C. M. Hall and M. Johnston (eds.) *Polar Tourism,* New York: John Wiley.

Stonehouse, B. and Snyder, J. (2007) 'Polar Tourism in Changing Environments', in J. Snyder & B. Stonehouse (eds.) *Prospects for Polar Tourism,* Wallingford: CABI.

Struck, G. (2004) 'Endangered wonders: Places to see before they vanish', *Condé Nast Traveller,* May.

Tickell, O. (2007) 'Tourism threatens Antarctica', *The Sunday Times.* Online: www.timesonline.co.uk/tol/travel/holiday_type/cruises/article1886800.ece (accessed 1 January 2011).

Tin, T., Fleminga, Z. L., Hughes, K. A., Ainley, D. G., Convey, P., Moreno, C. A., Pfeiffer, S., Scott, J. and Snape, I. (2008) 'Impacts of local human activities on the Antarctic environment', *Antarctic Science* 21: 3–33.

Tracey, P. J. (2001) 'Managing Antarctic Tourism', unpublished PhD thesis, University of Tasmania.

Turner, J., Colwell, S. R., Marchall, G., Lachlan-Cope, T., Carleton, A., Jones, P., Lagun, V., Reid, A. and Iagovkina, S. (2005) 'Antarctic climate change during the last 50 years', *International Journal of Climatology* 25: 279–94.

Vaughan, D. G., Marshall, G. J., Connolley, W. M., Parkinson, C., Mulvaney, R., Hodgson, D. A., King, J. C., Pudsey, C. J. and Turner, J. (2003) 'Recent rapid regional climate warming on the Antarctic Peninsula', *Climatic Change* 60: 243–74.

Vereda, M. (2010) 'Images of Antarctica and Ushuaia (Argentina) as a Gateway Port', in C. M. Hall and J. Saarinen (eds.) *Tourism and change in polar regions: climate, environment and experience,* Abingdon, UK: Routledge.

Wiley, J. (2008) '10 places to go before global warming hits hard', *Seattle Times.* Online: seattletimes.nwsource.com/html/travel/2004419854_webwarmingtrips18.html (accessed 10 January 2011).

# 4 Last chance tourism in alpine regions

## Last chance to ski?

*Robert Steiger, Jackie Dawson, and Johann Stötter*

### Introduction

The latest report released by the Intergovernmental Panel on Climate Change (IPCC) stated, 'there is new and stronger evidence of observed impacts of climate change on unique and vulnerable systems . . . such as . . . high mountain communities and ecosystems' (Meehl *et al.* 2007: 64). Given that mountain regions are highly susceptible to changing climatic conditions, in particular changing temperature regimes, a significant amount of research attention has been placed on identifying potential impacts and future outcomes for communities and businesses that rely on alpine environments. In particular the alpine ski industry has received detailed attention because of the sector's high cultural and economic importance in many regions (see Scott, Dawson and Jones 2008). Scott and McBoyle (2007) estimate the international ski industry to be worth over nine billion US dollars annually. In the United States, the National Ski Areas Association (NSAA) reported revenues of almost US$3.3 billion per year (NSAA, 2006), western Europe and Japan each reported revenues of over US$3 billion (Lazard 2002), and Australia reported their ski industry to be worth US$1.3 billion annually (National Institute of Economic and Industry Research 2006).

The multi-billion dollar ski sector is highly vulnerable to changes in both regional and local climate and has been projected to experience decreased natural snow reliability resulting in decreased season length, increased snow-making requirements, increased operating costs, decreased revenues in association with decreased visitation, and the closure of ski area operations (Abegg *et al.* 2007; Scott *et al.* 2008; Dawson and Scott 2010). Variable climate conditions experienced in the past have already influenced the closure of many ski areas internationally. For example, the consecutively poor snow seasons experienced in North America through the late 1980s in part caused the closure of hundreds of ski resorts across the US Northeast region (see NELSAP 2010). Simultaneously ski resorts in the European Alps experienced a series of snow-scarce winters near the end of the 1980s and into the 1990s. During this time Switzerland experienced average revenue losses of 20 per cent (1989–90) (Abegg 1996), and South Tyrol (Italy) ski lift transports were reduced by one-third (1988–89 and 1989–90) (Steiger 2011a). This trend towards ski area closures and ski resort bankruptcy is projected to continue as ski businesses become unable to afford the cost of adapting to newer and warmer climate systems (i.e. by investing in high efficiency snow-making technology, moving to higher elevations, diversifying product offerings such as snow tubing, retail, pools, bars, etc.) (see Scott and McBoyle 2007). If these projections hold true, it is possible that 'last chance tourism' (LCT) may become a reality for some highly vulnerable alpine ski regions before the end of the twenty-first century. But because LCT is about perceptions and responses of individuals who value or desire to experience a particular resource that is at risk (see Lemelin *et al.* 2010), the future implications of expected bio-physical impacts will depend heavily on the behavioural responses of individuals. For example, it is important to try to understand how tourists might react if they knew or felt that it would be their last chance to ski at a particular resort or in a certain geographic region. A common behavioural response to changed recreational conditions (i.e. such as the closure of a local ski area) involves substituting the activity that is no longer available for an all-together different activity, changing the timing of participation, or travelling to a different location in order to continue to participate (Iso-Ahola 1986). Exactly how individuals react to last chances to ski will play an important role in the overall vulnerability of the sector and will also aid policy and decision makers in the development and implementation of adaptive strategies and plans. This chapter discusses the implications of LCT for the alpine ski sector by outlining the relative vulnerability of the industry and discussing expected behavioural responses of ski tourists to 'last chances to ski'.

## Climate change vulnerability of alpine ski tourism

Projected climate warming is expected to impact both summer and winter skiing opportunities. The loss of ice caps and glaciers used for summer skiing in Europe has rapidly increased since the early 1990s (Allison *et al.* 2009). Alpine glaciers in Europe are projected to decline by 75 per cent (+/–15 per cent depending on the global climate model used) by the end of the twenty-first century (Radic and Hock 2011); their retreat will negatively influence summer ski tourism (Müller and Weber 2008; OcCC 2008). Summer glacier skiing became popular in the 1970s (Abegg *et al.* 1994) when most of the glacier ski areas were developed. In the 1990s, the summer skiing market in Switzerland was already saturated (Abegg *et al.* 1994), which influenced a declining number of summer skiers in all ski areas except the largest ones (i.e. Zermatt and Saas Fee).

Due to the fact that most of the glaciers used for skiing are relatively small and as a result are highly vulnerable to climate change, there have been significant changes over the past 20 years (Figure 4.1). In Austria, the maximum vertical drop of glacier ski slopes was reduced from 600 m in the 1990s to 300 m or less in the twenty-first century (e.g. Hintertux, Kitzsteinhorn). Between 1969 and 1998, the loss in total ice mass at all Austrian glaciers was estimated to be nearly 20 per cent (Kuhn *et al.* 2008) while their loss in surface area was slightly less at 17 per cent (Lamprecht and Kuhn 2007). Between 2001 and 2010 one of the biggest glaciers in Austria (Hintereisferner, with ca. 7.5 $km^2$) lost 20 per cent

*Figure 4.1* Ski lifts and glacier surface area in 1982/2009 at Kitzsteinhorn/Austria (Stahl, 2010)

of its volume (Bollmann *et al.* 2011). Glacier loss and retreat is accelerating with recent and recurring warming events causing significant damage to European glaciers. Haeberli and Maisch (2007) and Zemp *et al.* (2006) project that not more than 10 per cent of the Alpine glacier area will be left in the second half of the twenty-first century (also see Haeberli *et al.* 2007).

The shrinking glacier surfaces have caused access problems, considering ski lift stations were originally installed on the glacier or in bedrock near the glaciers. Due to glacial recession these lifts are now positioned several hundreds of meters away or even above the current glacier surface. Melting permafrost is also causing more frequent and intensified rock fall events and has destabilized infrastructure and foundations (e.g. ski lift pillars, restaurants), which has influenced a significant increase in maintenance costs. Glacier ski operations across Europe have already begun to disappear. In 1985 there were 32 resorts in total, in 1995 this dropped to 25, ten years later it was down to eight (ADAC 1985, 1995, 2010).

Abegg *et al.* (1994) examined the future consequences of projected glacier loss for Swiss ski areas finding that the expected upward shift of the equilibrium line by 100 m (caused by a +0.6°C scenario which is expected for the 2010s and 2020s, depending on the emission scenario) would result in a 50 per cent reduction of glacier surface area. A 200 m shift (caused by a +1.2°C scenario which is expected for the 2030s/2050s) would cause melting of more than 80 per cent of the glaciated area in current summer skiing areas (except large highly elevated glaciers such as Saas Fee and Zermatt in Switzerland). Under future warming conditions it is likely that demand for glacier skiing in Europe will be concentrated on the highly elevated Saas Fee and Zermatt while all others ski areas are expected to cease operation.

Winter ski resorts are also under significant strain resulting from changing climate conditions. There are now a number of peer-reviewed scholarly studies examining the impact that climate change is expected to have on ski resorts globally (Table 4.1). In order to understand the potential impact of climate change on ski resorts internationally Scott *et al.* (2003) developed an effective tool for modelling change in individual ski areas. The *ski sim I* and *ski sim II* models are used to project future ski season conditions (note: ski sim II is adapted for use in high alpine terrain and additionally accounts for spatially and temporally differing temperature lapse rates, more complex operational decisions for snow making and distributing climate variables to 100 m altitudinal bands to model snow depth in the entire ski area). The models utilize a combination of future climate change scenarios derived from General Circulation Model (GCM) outputs within desired time periods (i.e. 2020s, 2050s), with technical snow melt equations in order to determine future ski season length, snow-making requirements, and probability of being operational during economically important holiday periods. The models account for relative adaptive abilities of ski resorts including snow-making capacity through inclusion of certain technical capacities (i.e. minimum temperature at which snow can be made economically and daily snow-making capacity) and stakeholder derived decision-making rules (i.e. season start/end dates, target snow pack depth to maintain) (see Scott *et al.* 2003, 2006a). The models also allow for elevation adjustments thus reflecting the differing altitudes, and resulting temperature differences for each modelled ski area (see Dawson and Scott 2007, 2010). Outputs from these models can be used to

*Table 4.1* Climate change and ski tourism literature

| | |
|---|---|
| Australia | Bicknell and McManus 2006; Galloway 1988; Hennessy *et al.* 2003, 2008; König 1998, 1999 |
| Austria | Abegg *et al.* 2007; Breiling and Pavel 1999; Lambrecht and Kuhn 2007; Steiger and Mayer 2008; Steiger 2010, 2011b, 2011c; Töglhofer *et al.* 2011; Unbehaun *et al.* 2008; Wolfsegger *et al.* 2008 |
| Canada | McBoyle and Wall 1987, 1992; Lamothe and Périard Consultants 1988; Scott *et al.* 2003, 2006a, 2006b, 2008; Scott and McBoyle 2007; Shih *et al.* 2008 |
| France | Abegg *et al.* 2007 |
| Germany | Abegg *et al.* 2007 |
| Italy | Abegg *et al.* 2007; Steiger 2011a, 2011c |
| Japan | Fukushima *et al.* 2003 |
| Sweden | Moen and Fredman 2007 |
| Switzerland | Abegg 1996; Abegg *et al.* 2007, 2008; König and Abegg 1997; Behringer *et al.* 2000; Elsasser and Messerli 2001; Elsasser and Bürki 2002; Hoffmann *et al.* 2009; Teich *et al.* 2007; Uhlmann *et al.* 2008 |
| United States | Lipski and McBoyle 1991; Hayhoe *et al.* 2008; Casola *et al.* 2005; Dawson and Scott 2007; Hamilton *et al.* 2007; Hayhoe *et al.* 2008; Scott *et al.* 2008; Dawson *et al.* 2009 |

determine which ski resorts in an area of interest will be economically viable under future climate change conditions and which resorts are expected to cease operation. The extent to which individual ski areas are considered economically vulnerable is determined using two factors: 75 per cent probability of being operational during the economically vital Christmas–New Year holiday (i.e. open for full period seven and a half years out of ten), and the '100-day rule' (i.e. minimum of 100-day season); both indicators have been suggested to reflect profitability in Europe and North America (König and Abegg 1997; Bürki 2002; Scott *et al.* 2003, 2006a).

To understand the regional implications of climate change for the alpine ski sector Steiger (2010) and Dawson and Scott (2010) have used *Ski Sim II* and *I* respectively to examine ski area vulnerability in two geographic areas including the Tyrol region of Austria and Italy ($n = 111$ ski areas) and the Northeastern United States ($n = 103$ ski areas). Applying the 100-day rule in Tyrol, 77 per cent of the 111 ski areas would remain snow reliable in the 2050s timeframe under and A1B mid-range emissions scenario. However, this number drops to just 15 per cent in the 2080s timeframe (Figure 4.2). In the Tyrol region, the Christmas holidays last for two weeks and accounts for about 30 per cent of lift ticket sales (Steiger 2010). Thus the second indicator of economic viability is essential in understanding overall impacts. With this indicator the share of snow-reliable ski areas is lower than when using the 100-days rule (Figure 4.2), because the number of snow-making hours prior to that early date required to produce a sufficient snow cover is quite sensitive to increasing temperatures. Under the same A1B mid-range emissions scenario, 77 per cent of ski areas would remain snow reliable until the 2030s time period. This proportion of viable ski areas rapidly declines to 41 per cent in the 2050s and to 5 per cent in the 2080s. Under future climate change conditions viable ski areas in Tyrol are likely to concentrate in the central part of the study area, while ski areas at the northern and southern boundary are projected to disappear (Figure 4.2).

In the US Northeast region, Dawson and Scott (2010) found that during the 2020s time period only 55 and 54 per cent of the 103 currently operating ski areas were projected to be able to maintain 100-day season lengths under low (B1) and high (A1fi) emissions scenarios respectively. These figures drop to 54 and 40 per cent in the 2050s time period and 45 and 29 per cent in the 2080s time period respectively. At the state level, ski areas in the more southern states of Connecticut and Massachusetts are the most vulnerable as they are not expected to maintain season lengths of at least 100 days as early as the 2020s time period. Ski areas in Vermont are the least vulnerable to climate change, with 100 (B1) and 94 per cent (A1fi) of ski areas maintaining ski season lengths of greater than 100 days even into the 2080s time period (Table 4.2).

Using the second benchmark for economic viability (i.e. ski areas' ability to maintain a 75 per cent probability of being operational during the economically important Christmas–New Year holiday period), only 43 (B1 low emission) and 44 per cent (A1fi high emissions) were projected to maintain more than a 75 per cent probability for the 2020s timeframe. This projection decreased to 35 and

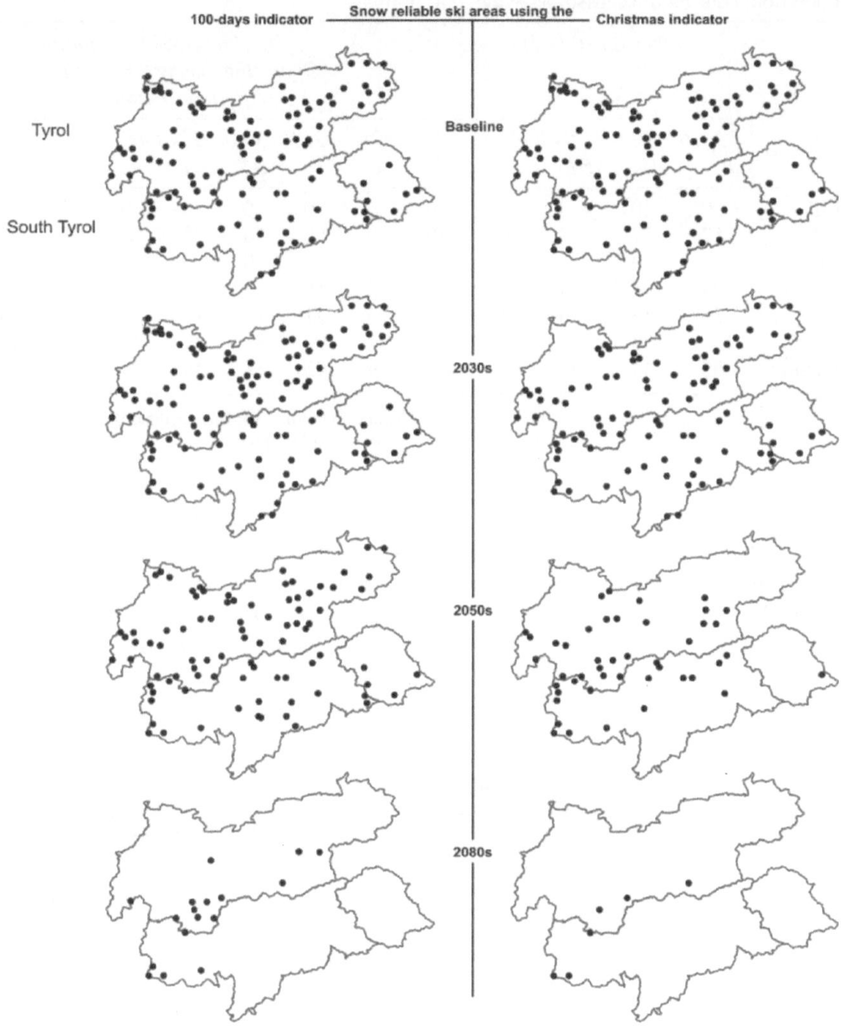


*Figure 4.2* Number of snow reliable ski areas in the Tyro–South Tyrol region, Austria, Italy (A1B scenario)

26 per cent respectively during the 2050s time period and to 33 and 7 per cent in the 2080s timeframe (Table 4.2). Over 20 per cent of ski area revenue is generated during the Christmas–New Year holiday period in the US Northeast regions, which occurs at the highly vulnerable beginning of the ski season (NSAA, 2007).

Examination of the two economic indicators (100-day season and >75 per cent probability of being operational during the Christmas–New Year period) revealed that ski area supply in the US Northeast is likely to experience a northward contraction as particular ski areas located in less elevated southern regions become

*Table 4.2* Per cent of ski areas with 100+ day season length and > 75% probability of being operational during the Christmas–New Year holiday

| | *% with 100+ Day Season* | | | | | | *% with >75% Probability Operating during Christmas-New Year holiday* | | | | | |
|---|---|---|---|---|---|---|---|---|---|---|---|---|
| | *2010–39* | | *2040–69* | | *2070–99* | | *2010–39* | | *2040–69* | | *2070–99* | |
| | *B1* | *A1* | *B1* | *A1* | *B1* | *A1* | *B1* | *A1* | *B1* | *A1* | *B1* | *A1* |
| CT | 0 | 0 | 0 | 0 | 0 | 0 | 0 | 0 | 0 | 0 | 0 | 0 |
| ME | 57 | 57 | 57 | 50 | 50 | 50 | 50 | 50 | 50 | 50 | 50 | 50 |
| MA | 8 | 0 | 0 | 0 | 0 | 0 | 0 | 0 | 0 | 0 | 0 | 0 |
| NH | 94 | 94 | 94 | 50 | 67 | 33 | 50 | 33 | 39 | 17 | 33 | 0 |
| NY | 36 | 36 | 36 | 22 | 25 | 6 | 31 | 17 | 17 | 3 | 14 | 0 |
| VT | 100 | 100 | 100 | 94 | 100 | 94 | 94 | 89 | 89 | 83 | 89 | 22 |
| **Region** | **55** | **54** | **54** | **40** | **45** | **29** | **43** | **34** | **35** | **26** | **33** | **7** |

Total number of ski areas: Connecticut (CT) n = 5, Maine (ME) n = 14, Massachusetts (MA) n = 12, New Hampshire (NH) n = 18, New York (NY) n = 36, Vermont (VT) n = 18

unable to afford the increased cost of operating under projected climate change conditions. Of the 103 operating ski areas, 41 are expected to remain economically sustainable by the 2020s time period under the high (A1fi) emission scenario. The number of operational ski areas drops to 34 for the 2050s time period and 30 for the 2080s time period (Figure 4.3). Under a lower (B1) emissions

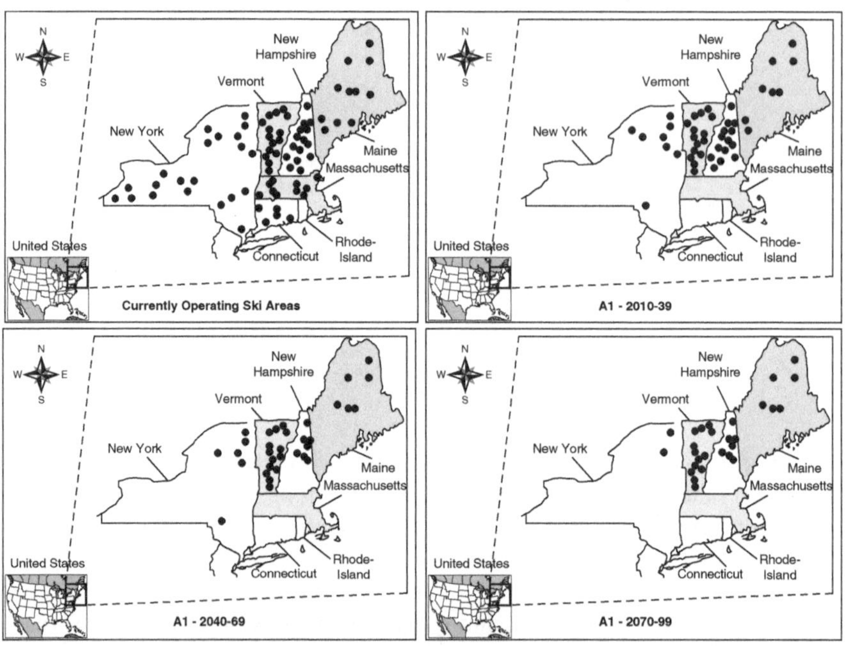


*Figure 4.3* Number of snow reliable ski areas in the US Northeast region (A1 scenario)

scenario the number of operating ski areas in the 2020s time period is higher at 42. This number drops to 41 and further to 35 during the 2050s and 2080s time periods respectively.

## Behavioural response to emerging 'last chances to ski'

Previous studies, including those discussed above, recognize that it is not the entire ski sector that is at risk to climate change but rather at risk are certain individual ski areas that collectively make up a particular ski marketplace (Abegg *et al.* 2007; Scott *et al.* 2008). Also at risk are local communities and residents that rely on alpine tourism to drive local economies and provide employment. Despite understanding that future climate change will present many challenges relating to the closure of certain ski areas internationally, very little is known about how individual tourists might respond to the expected changes in ski resort supply and operation. For example, it is unclear how tourists might react if they knew it would be their last chance to ski at a particular resort or in a certain geographic region. This is of concern considering how easily tourists can adapt their behaviour in response to climate variability, poor snow conditions, or ski resort closures – especially in comparison to the difficulty and expense involved in structural and management-based adaptations currently being used or considered by ski areas (see Scott and McBoyle 2007). Because of this ease, it is surprising that so few studies have examined the impact that climate change is expected to have on the response of individuals including the emerging notions surrounding LCT (see Dawson, Havitz and Scott, 2011 for a list of demand-side studies).

A common behavioural response to changed recreational conditions (i.e. such as the closure of a local ski area) involves substituting the activity that is no longer available for an all-together different activity. This is known as activity substitution as defined within the 'theory of substitutability' developed by Iso-Ahola (1986). Other substitution responses involve changing the amount of time one participates in the activity (i.e. skiing less because of poor conditions) or substituting the place where the activity usually takes place (i.e. destination substitution – see Stewart *et al.* this volume). These behavioural substitution options represent possible demand-side reactions to last chances to ski. Understanding exactly how individuals might react to last chances in the alpine ski sector will play an important role in understanding the overall vulnerability of the sector and will also help local communities and residents develop and implement appropriate adaptation plans to deal with changing conditions.

Both König (1998) and Dawson and Scott (2010) examined the extent to which individuals might engage in skier substitution behaviour resulting from climate variability finding that very few respondents would stop alpine skiing (6 per cent), but rather they were more likely to travel long distances to ski at another more viable resort (38 per cent). This outcome is unlike some other LCT situations where an entire landscape, species or activity is expected to disappear completely; alpine skiing opportunities will only be lost within certain geographical regions while other regions will remain entirely operational. Revealed behavioural substitution intentions amongst alpine skiers suggest that demand for skiing opportunities is likely to remain stable or even diminish proportionally less than the expected decrease in supply (i.e. ski area closures). This is likely to result in a net transfer of demand throughout the marketplace that remains operational – meaning that while some communities, business and local residents will be negatively impacted by ski area closures, others stand to benefit.

Within the US Northeast, Dawson and Scott (2010) suggest that by the 2050s time period there will be a contraction of the ski industry from south to north favouring certain individual ski areas in Vermont, eastern New York, Maine and New Hampshire over those expected to cease operation in western New York, Massachusetts and Connecticut (see Figure 4.2; also see Scott *et al.* 2008). Nationally, the US Northeast could benefit from ski area closures in the US Midwest region, but the market share could be also split with the US Rocky Mountain region (northwest coast) and Canada, which are expected to fair well under future climate change scenarios (see Dawson *et al.* 2009).

## Conclusion

Resorts that are projected to cease operation due to economic pressures associated with changing climate and other conditions may, to some extent, experience a last chance to ski phenomenon this century. However, it will be unlike the outcomes expected for other LCT destinations where entire landscapes, wildlife populations, or cultures may disappear. LCT within the alpine ski sector is likely to create a situation where the loss of some ski business will benefit existing viable ski areas due to an expected market shift in demand (i.e. demand is not expected to decrease proportionally to projected decrease in supply due to expected substitution behaviour – see above; also see Dawson *et al.* 2011).

Under this scenario communities and local residents who rely on extremely vulnerable ski areas will need to prepare for ski resort closures and consider economic restructuring options. Simultaneously, the communities and ski businesses projected to remain operational, and thus likely to experience an increase in market share (i.e. demand), will need to prepare for future change including expected revenue increases as well as development pressures. Steiger (2010) notes that under this scenario increased pressure can be expected within the Tyrol–South Tyrol region, associated with negative impacts of increased visitation on alpine ecology. Other pressures related to the concentration of ski tourism in fewer areas include increased water use for snowmaking, real-estate development, slope expansion and congestion. Transportation infrastructure, including ski area parking lots and roadways may require expansion with increased visitation and congestion. Community-based impacts could include increased pressure on local services and environmental resources as well as increasing real-estate values, so much so that in some cases local residents may no longer be able to afford to live nearby.

Currently, neither ski areas in the European Alps or North America promote ski tourism as a last chance event. This is likely because operators want to sustain the image of 'absolute snow guarantee' and because only certain alpine skiing regions are actually at risk of disappearing altogether. However, a number of resorts, including the remaining summer skiing resorts in Europe are beginning to highlight their destination as being one of the few remaining summer skiing resorts left (e.g. see the webpages of Hintertuxer Gletscher or Zermatt). These resorts – especially those in Austria – have tried to strengthen their summer alpine products by more notably promoting and staging alpine glaciers for summer viewing (e.g. the 'Dachstein Ice Palace' at Dachsteiner Gletscher, 'Nature's Ice Palace' at Hintertuxer Gletscher) (see Der Dachstein 2011; Hintertuxer Gletscher 2011). A positive trend in summer guests after the record warm summer season 2003 can also be interpreted as some anecdotal evidence that there may be a desire to see glaciers before they disappear.

In the upcoming decades, skiing is likely to become an elitist holiday activity as prices considerably increase due to higher operational costs (i.e. snow making) and as competition decreases due to fewer operating ski areas. Despite the notion that alpine skiing will become an elitist activity and that the sector will experience a significant contraction as vulnerable ski areas cease operation due to climate change, it should not yet be considered a last chance tourism product. However, as more and more ski areas close over the course of the twenty-first century there will be many situations where individuals will experience or perceive a last chance to ski phenomenon, particularly at highly vulnerable resorts or within vulnerable geographical regions.

## References

Abegg, B. (1996) *Klimaänderung und Tourismus. Klimafolgenforschung am Beispiel des Wintertourismus in den Schweizer Alpen. Zurich*. Zurich: vdf.

Abegg, B., König, U., and Maisch, M. (1994) Klimaänderung und Gletscherskitourismus', *Geographica Helvetica* 3: 103–114.

Abegg, B., Agrawala, S., Crick, F., and de Montfalcon, A. (2007) 'Climate change impacts and adaptation in winter tourism' (pp. 25–60) in S. Agrawala (ed.) *Climate Change in the European Alps. Adapting Winter Tourism and Natural Hazards Management*. Paris: OECD.

Abegg, B., Kolb, M., Sprengel, D., and Hoffmann, V.H. (2008) 'Klimawandel aus der Sicht der Schweizer Seilbahnunternehmer' (pp. 73–83) in T. Bieger, C. Laesser, and R. Maggi (eds.) *Jahrbuch der Schweizerischen Tourismuswirtschaft*. St. Gallen: IDT-HSG.

ADAC (1985) *Der große ADAC Ski Atlas*, Munich: ADAC Verlag.

ADAC (1995) *Der ADAC Ski Atlas Alpen*, Munich: ADAC Verlag GmbH.

ADAC (2010) *ADAC SkiGuide 2010*, Munich: ADAC Verlag GmbH.

Allison, I., Bindoff, N.L., Bindschadler, R.A., Cox, P.M., Noblet, N., England, M.H., *et al.* (2009) *The Copenhagen Diagnosis. Updating the World on the Latest Climate Science*. Sydney, Australia.

Behringer, J., Bürki, R., and Fuhrer, J. (2000) 'Participatory integrated assessment of adaptation to climate change in alpine tourism and mountain agriculture', *Integrated Assessment* 1(3): 31–338.

Bicknell, S., and McManus, P. (2006) 'The canary in the coalmine: Australian ski resorts and their response to climate change', *Geographical Research* 44: 386–400.

Bollmann, E., Sailer, R., Briese, C., Stötter, J., and Fritzmann, P. (2011) 'Potential of airborne laser scanning for geomorphologic feature and process detection and quantifications in high alpine mountains', *Geomorphol* 54: 566–571.

Breiling, M., and Pavel, C. (1999) 'The impact of global warming on winter tourism and skiing: A regionalised model for Austrian snow conditions', *Regional Environmental Change* 1 (1): 4–14.

Casola, J., Kay, J., Snover, A., Norheim, R., and Binder, L. (2005) *Climate Impacts on Washington's Hydropower, Water Supply, Forests, Fish and Agriculture*. Washington, DC: University of Washington, Centre for Science and the Earth System, Seattle.

Dawson, J., and Scott, D. (2007) 'Climate change vulnerability of the Vermont ski tourism industry (USA)', *Annals of Leisure Research* 10 (3&4): 550–572.

Dawson, J., and Scott, D. (2010) 'Systems analysis of climate change vulnerability for the US northeast ski sector', *Tourism and Hospitality Planning & Development* 7(3): 219–235.

Dawson, J., Scott, D., and McBoyle, G. (2009). 'Analogue analysis of climate change vulnerability in the US Northeast ski tourism', *Climate Research* 39(1): 1–9.

Dawson, J., Havitz, M.E., and Scott, D. (2011) 'Behavioural adaptation of alpine skiers to climate change: examining activity involvement and place loyalty', *Journal of Travel and Tourism Marketing* 28(4): 388–404.

Der Dachstein (2011) 'Dachstein Ice Palace'. Online: www.derdachstein.at/en/summer/Dachstein_Ice_Palace.html (accessed 8 April 2011).

Elsasser, H., and Messerli, P. (2001) 'The Vulnerability of the snow industry in the Swiss Alps', *Mountain Research and Development* 21(4): 335–339.

Elsasser, H., and Bürki, R. (2002) 'Climate change as a threat to tourism in the Alps', *Climate Research* 20: 253–257.

Erickson, J. (2005) 'Changes in air, part 3: bleak forecast for the ski industry', *Rocky Mountain News*, 19 March. Online: http://rockymountainnews.com (accessed 25 March 2010).

Fukushima, T., Kureha, M., Ozaki, N., Fukimori, Y., and Harasawa, H. (2003) 'Influences of air temperature change on leisure industries: case study on ski activities', *Mitigation and Adaptation Strategies for Climate Change* 7(2): 173–189.

Galloway, R. (1988) 'The potential impact of climate changes on Australian ski fields' (pp. 428–437) in G. Pearman (ed.) *Greenhouse Planning for Climate Change*. Melbourne: CSIRO publications.

Haeberli, W., and Maisch, M. (2007) 'Klimawandel im Hochgebirge' (pp. 98–107) in W. Endlicher and F.W. Gerstengarbe (eds.) *Der Klimawandel – Einblicke, Rückblicke und Ausblicke*. Zürich: University of Zürich.

Haeberli, W., Hoelzle, M., Paul, F., and Zemp, M. (2007) 'Integrated monitoring of mountain glaciers as key indicators of global climate change: the European Alps', *Annals of Glaciology* 46: 150–160.

Hamilton, L., Brown, C., and Keim, B. (2007) 'Ski areas, weather and climate: time series models for New England case studies', *International Journal of Climatology* 27: 2213–2124.

Hayhoe, K., Wake C., Anderson B., Liang, X-Z., Maurer E., Zhu, J., Bradbury J., DeGaetano, A., Stoner A.M. and Wuebblses, D. (2008) 'Regional climate change projections for the Northeast USA', *Mitigation Adaptationg Strategies for Global Change* 13: 425–436.

Hennessy, K., Whetton, P., Smith, I., Bathols, J., Hutchinson, M., and Sharples, J. (2003) *The Impact of Climate Change on Snow Conditions in Mainland Australia.* Hobart: CSIRO Marine and Atmospheric Research.

Hintertuxer Gletscher (2011). ‚Nature's ice palace'. Online: www.hintertuxergletscher.at/en/news/0/Natureispalastenglisch.html (accessed 8 April 2011).

Hoffmann, V.H., Sprengel, D., Ziegler, A., Kolb, M., and Abegg, B. (2009) 'Determinants of corporate adaptation to climate change in winter tourism: An econometric analysis', *Global Environmental Change* 19(2): 256–264.

Iso-Ahola, S.E. (1986) 'A theory of substitutability of leisure behavior', *Leisure Sciences* 8(4): 367–389.

König, U. (1998) *Tourism in a Warmer World: Implications of Climate Change due to enhanced Greenhouse Effect for the Ski Industry in the Australian Alps*, Zurich: University of Zurich, Wirtschaftsgeographie und Raumplanung.

König, U. (1999) 'Climate Change and Snow Tourism in Australia', *Geographica Helvetica* 54 (3): 147–157.

König, U., and Abegg, B. (1997) 'Impacts of climate change on tourism in the Swiss Alps', *Journal of Sustainable Tourism* 5(1): 46–58.

Kuhn, M., Lambrecht, A., Abermann, J., Patzelt, G., and Gross, G. (2008) *Die österreichischen Gletscher 1998 und 1969*. Vienna: Flächen- und Volumensänderungen.

Lambrecht, A., and Kuhn, M. (2007) 'Glacier changes in the Austrian Alps during the last three decades, derived from the new Austrian glacier inventory', *Annals of Glaciology* 46: 177–184.

Lamothe and Periard Consultants (1988) *Implications of Climate Change for Downhill Skiing in Québec*. Toronto: Meteorological Service of Canada.

Lazard, A. (2002) 'Ski winter: world flat', Ski Area Management, 24–7 September.

Lemelin, H., Dawson, J., Stewart, E.J., Maher, P., and Lück, M. (2010) 'Last-chance tourism: the boom, doom, and gloom of visiting vanishing destinations', *Current Issues in Tourism* 13: 477–493.

Lipski, S., and McBoyle, G. (1991) 'The impact of global warming on downhill skiing in Michigan', *East Lakes Geographer* 26: 37–51.

McBoyle, G., and Wall, G. (1987) 'The impact of $CO_2$ induced warming on downhill skiing in the Laurentians', *Cahiers de géographie du Québec* 31(82): 39–50.

McBoyle, G., and Wall, G. (1992) 'Great lakes skiing and climate change' (pp. 70–81), in A. Gill and R. Hartmann (eds.) *Mountain Resort Development*. Burnaby: SFU.

Meehl, G.A., Stocker, T.F., Collins, W.D., Friedlingstein, P., Gaye, A.T., Gregory, J.M., Kitoh, A., Kuntti, R., Murphy, J.M., Noda, A., Raper, S.C.B., Watterson, I.G., Weaver, A.J., and Zhao, Z.C. (2007) 'Global Climate Projections', in S. Solomon, D. Qin, M. Manning, Z. Chen, M. Marquis, K. B. Averyt, M.Tignor, H. L. Miller (eds.) *Climate Change 2007: The Physical Science Basis. Contribution of Working Group I to the Fourth Assessment Report.* Cambridge: Intergovernmental Panel Climate Change.

Moen, J., and Fredman, P. (2007) 'Effects of climate change on alpine skiing in Sweden', *Journal of Sustainable Tourism* 15(4): 418–437.

Müller, H., and Weber, F. (2008) 'Climate change and tourism – scenario analysis for the Bernese Oberland in 2030', *Tourism Review* 3: 57–71.

National Institute of Economic and Industry Research (2006) *The economic significance of the Australian alpine resorts*. Melbournelia: Alpine Resorts Coordinating Council.

NELSAP (2010) 'New England's Lost Ski Areas Project'. Online: www.nelsap.org/ (accessed 20 May 2010).

NSAA (2006) *2005–06 Economic analysis of United States ski areas*. Lakewood, CO: National Ski Areas Association.

NSAA (2007) *Kottke National End of Season Survey 2006/07*, 28th edn. Lakewood, CO: National Ski Areas Association.

OcCC (2008) *Das Klima ändert – was nun? Der neue UN-Klimabericht (IPCC 2007) und die wichtigsten Ergebnisse aus Sicht der Schweiz.* Bern: OcCC.

Radic, V., and Hock, R. (2011) 'Regionally differentiated contribution of mountain glaciers and ice caps to future sea-level rise', *Nature Geoscience* 4(2): 91–94.

Scott, D., and McBoyle, G. (2007) 'Climate change adaptation in the ski industry', *Mitigation and Adaptation Strategies for Global Change* 12: 1411–1431.

Scott, D., McBoyle, G., and Mills, B. (2003) 'Climate change and the skiing industry in southern Ontario (Canada): exploring the importance of snowmaking as a technical adaptation', *Climate Research* 23: 171–181.

Scott, D., McBoyle, G., Mills, B., and Minogue, A. (2006a) 'Climate change and sustainability of ski-based tourism in eastern North America: a reassessment', *Journal of Sustainable Tourism* 14(4): 376–398.

Scott, D., McBoyle, G., and Minogue, A. (2006b) 'Climate Change and Quebec′s Ski Industry', *Global Environmental Change* 17: 181–190.

Scott, D., Dawson, J., and Jones, B. (2008) 'Climate change vulnerability of the US Northeast winter recreation– tourism sector', *Mitigation and Adaptation Strategies for Global Change* (13): 577–596.

Shih, C., Nicholls, S., and Holecek, D.F. (2008) 'Impact of Weather on Downhill Ski Lift Ticket Sales', *Journal of Travel Research* 47(3): 359–372

Stahl, T. (2010) Die Erweiterung der Beschneiungsanlage am Kitzsteinhorn aus betriebswirtschaftlicher Sicht. Online: www.stahl.co.at/wp-content/uploads/2010/05/Die-Erweiterung-der-Beschneiungsanlage-am-Kitzsteinhorn-aus-betriebswirtschaftlicher-Sicht.pdf (accessed 24 April 2011).

Steiger, R. (2010) 'The impact of climate change on ski season length and snowmaking requirements', *Climate Research* 43(3): 251–262.

Steiger, R. (2011a) 'The impact of climate change on ski touristic demand using an analogue Approach', in K. Weiermair, H. Pechlahner, A. Strobl, M. Elmi, and M. Schuckert (eds.) *Coping with Global Climate Change: Strategies, Policies and Measures for the Tourism Industry*. Innsbruck: Innsbruck University Press (in press).

Steiger, R. (2011b) 'The impact of snow scarcity on ski tourism. An analysis of the record warm season 2006/07 in Tyrol (Austria)', in *Tourism Review* (in press).

Steiger, R. (2011c) 'Assessment of climate change impacts on winter tourism in Tyrol (Austria, Italy)', *Tourism Geographies* (in press).

Steiger, R., and Mayer, M. (2008) 'Snowmaking and climate change. future options for snow production in Tyrolean ski resorts', *Mountain Research and Development* 28(3/4): 292–298.

Teich, M., Lardelli, C., Bebi, P., Gallati, D., Kytzia, S., Pohl, M. *et al.* (2007) *Klimawandel und Wintertourismus: Ökonomische und ökologische Auswirkungen von technischer Beschneiung.* Birmensdorf/Davos.

Töglhofer, C., Eigner, F., and Prettenthaler, F. (2011) 'Climatic and economic impacts on tourism demand in Austrian ski areas', *Climate Research* 46(1): 1–14.

Uhlmann, B., Goyette, S., and Beniston, M. (2008) 'Sensitivity analysis of snow patterns in Swiss ski resorts to shifts in temperature, precipitation and humidity under conditions of climate change', *International Journal of Climatology* 29(8): 1048–1055.

Unbehaun, W., Probstl, U., and Haider, W. (2008) 'Trends in winter sports tourism: challenges for the future', *Tourism Review* 63(1): 36–47.

Wolfsegger, C., Gössling, S., and Scott, D.(2008) 'Climate change risk appraisal in the Austrian ski industry', *Tourism Review International* 12(1): 13–23.

Zemp, M., Haeberli, W., Hoelzle, M., and Paul, F. (2006) 'Alpine glaciers to disappear within decades?', *Geophysical Research Letters* 33(13).

# 5 Vanishing fauna of tourism interest

*David Newsome and Kate Rodger*

## Introduction

Wildlife tourism is continuing to grow as an industry resulting in an increased number of different wildlife watching activities developing around the world in a wide and diverse range of ecosystems (Fennell 2008; Newsome, Dowling, and Moore 2005; Rodger, Moore, and Newsome 2007). Human–wildlife experiences provide opportunities for people to observe, photograph and interact with animals that may be endangered, threatened or rare, and are increasingly being offered in a number of destinations around the world (Ballantyne, Packer, and Hughes 2009; Newsome *et al.* 2005; Shackley 1996).

In the book *100 Animals to See Before they Die*, Hilary Brandt states:

> If the animal you are observing is an endangered species it adds an emotional factor that you can never find in a zoo. The experience is unforgettable, even if you only glimpsed the animal. And to understand why its future is uncertain you need to see where it lives and appreciate – and worry about – the human pressures that have put it at risk. Then, once you've developed this emotional link, you will want to know how to help.
>
> (Garbutt and Unwin 2007: 3)

There is a wide range of species that fit into the rare and endangered category. Such species include the giant panda (*Ailuropoda melanoleuca*), various river dolphins (e.g. *Inia geoffrensis*), all species of rhinoceros (e.g. *Rhinoceros unicornis*), many species of birds and various reptiles such as the Komodo dragon (*Varanus komodensis*). Even the African elephant (*Loxodonta africanus* and *Loxodonta cyclotis*) is considered under threat as it has undergone a dramatic decline across Africa over the last 200 years (Garbutt and Unwin 2007). Perhaps one of the most enigmatic of all wildlife is the tiger (*Panthera tigris*), with only about 3,000 left in the wild, it is a species that has become a significant focus of domestic and international tourism activity in the national parks in India where it can be readily viewed (Figure 5.1).

Changes within the natural ecosystems have caused a number of wildlife species to now be considered to be rare and/or endangered. Reasons for these

*Figure 5.1* Tiger safari, Kanha National Park, India (Photograph courtesy of R. Chyne)

changes and the resulting decline in wildlife populations are long and varied with the most topical one being climate change. Climate change is a global issue and has been described as one of the biggest threats to society (Kachel and Jennings 2010). Wildlife tourism, similar to all tourism, can be seen as a contributor to and a 'victim' of climate change (WTO 2003).

In recent years public awareness of the impact of climate change through television documentaries and media coverage of endangered species, vanishing destinations and/or wildlife has increased (Curtin 2009). For example, in 1990 Douglas Adams and Mark Carwardine published a book called *Last Chance to See*. It described a series of journeys which the authors undertook to various places around the world to look for some of the rarest and most endangered animals. Some of the wildlife they viewed included the Komodo dragon of Indonesia (*Varanus komodensis*), the Kakapo (*Strigops habroptila*) of New Zealand, the blind river dolphins (*Platanista* spp.) of China, the white rhinoceros (*Ceratotherium simum*) and gorillas (*Gorilla beringei beringei*) in Zaire (now known as the Democratic Republic of Congo) (Adams and Carwardine 1990). It is well recognised that many wildlife species are disappearing and tourism provides the opportunity to view these species before they are gone for good. The outcome is an emerging new trend in wildlife tourism known as 'last chance wildlife tourism' (Lemelin *et al.* 2010). Last chance tourism, or 'doom tourism', can be defined as 'tourists explicitly seek vanishing landscapes or seascapes, and/or disappearing natural and/or social heritage' (Lemelin *et al.* 2010: 478). This includes the desire to see unique, rare, special, enigmatic and difficult to view wildlife species, some of which are quickly disappearing. This view may differ from that of the traditional wildlife tourist where the focus will be on wildlife per se rather than rare and endangered species. For the last chance wildlife tourist it is unclear how different the motivation (and awareness of environmental issues) to see specific wildlife might be. In this regard, although tour operators are emerging that market 'last chance to see' itineraries such tours also focus on other natural attributes such as waterfalls and river systems (see Easier Inc. 2011).

Tourism operators are increasingly marketing the 'last chance concept' especially when rare species can be viewed under natural conditions (for example, orangutanadventures.com, no date). It is, however, debatable whether publicising and promoting the vulnerability of threatened and/or rare species can help or hinder wildlife conservation. In fact, Courchamp *et al.* (2006) hypothesised that greater value is often placed on rare species resulting in a downward spiral because increased rarity can lead to increased exploitation and consequently a reduction in population size, due to detrimental impacts, which in turn leads to an even greater increase in value on the wildlife and greater demand (as in the case of the Kinabatangan River system in Sabah which is discussed later in this chapter). This spiral continues on and on with the final outcome potentially being the extinction of the species. This process they termed the 'anthropogenic allee effect'. Ecotourism activities have been shown to have negative impacts on wildlife (for example, see Newsome *et al.* 2005) and it is therefore likely that rare and/or threatened species will be disproportionately impacted upon by wildlife tourism activities (Courchamp *et al.* 2006; Hall *et al.* 2008).

Furthermore, threatened species and habitats often occur in countries where there are limited resources to protect them (Curtin 2009). Questions can be asked about whether tourism is a success story in helping people to appreciate and conserve species or whether in fact it will have the opposite effect. The objective here therefore is to explore the nature of tourism interest in vanishing fauna with a focus on selected rare primates and birds. Using Duffus and Dearden's (1990) conceptual framework of wildlife tourism, we will examine changes in the types and number of tourists seeking 'last chance' wildlife tourism interactions whilst highlighting the need to discuss what approach is best for the long-term sustainability of last chance wildlife tourism interactions as visitor pressure increases on the wildlife and their environment.

## Conceptual framework

Wildlife tourism activities and the resulting interactions are not static. They exhibit change and evolution in the users and the sites where the activity takes place (Duffus and Dearden 1990; Higham 1998). For wildlife tourism there appears to be an increased number of last chance tourists seeking to interact with rare and/or endangered species. This in turn results in an increased number of tourists and corresponding evolution of the site to cater for these wildlife tourism interactions. Duffus and Dearden (1990) explore three main ideas: the wildlife tourist, target wildlife and their habitats, and the historical relationships between them (Catlin *et al.* 2011). The emphasis here, however, is on understanding the nature of the tourism visitor situation as it changes from specialist wildlife viewing to generalist wildlife viewing. More specifically, Duffus and Dearden's (1990) conceptual framework for wildlife tourism (Figure 5.2) identified the changing nature of wildlife tourists and how the wildlife resources evolve and change over time in terms of both the sites where the activities take place (site evolution) and the users (site user evolution) (Higham 1998). Through time a site develops a public image due to the growth in publicity and facilities designed to service the visitors who arrive at a natural area to encounter wildlife. However, the expansion of facilities in turn influences the types of individuals who visit a site, visitor expectations, visitor satisfaction derived from the human–wildlife interaction and visitor impact on the wildlife (Duffus and Dearden 1990). Because last chance wildlife tourism

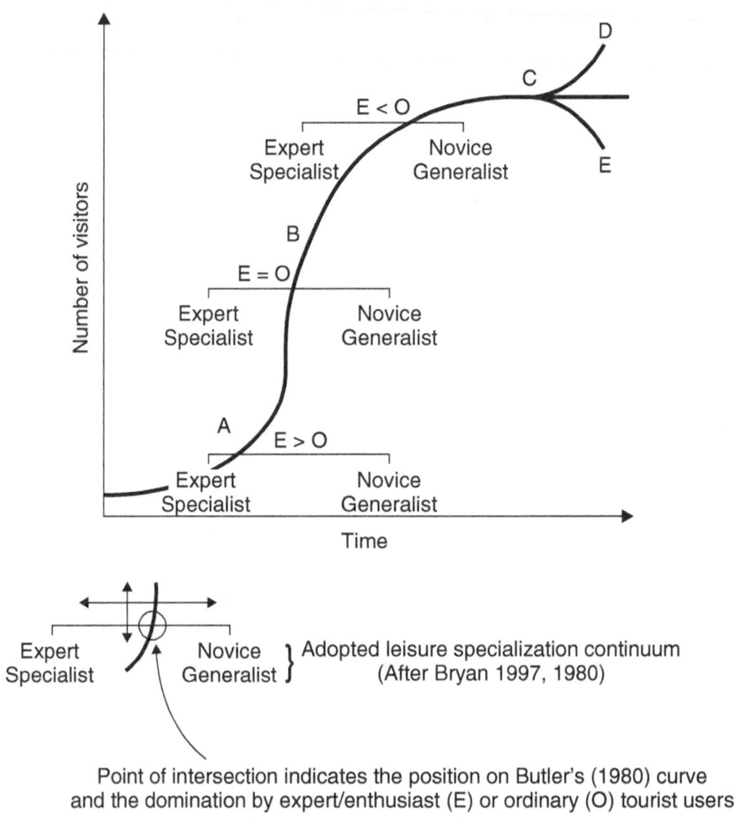


*Figure 5.2* User specialisation and site evolution (derived from Duffus and Dearden 1990)

activities can now be added to a long list of wildlife tourism categories (see Reynolds and Braithwaite, 2001), coupled with increasing tourism pressure, evolution of the visitor profile as described by Duffus and Dearden (1990) and Higham (1998) is likely to be an issue for many wildlife tourism destinations in the future.

Duffus and Dearden (1990) describe a sequential change in the type of tourist that can be seen beginning with exploratory users or specialists. Specialists are generally visitors who require minimal infrastructure, interpretive facilities, and on the whole their presence is absorbed by the existing social and ecological systems already in place. The specialist tourist is generally thought to be experienced and knowledgeable (Higham, Lusseau, and Hendry 2008). At this stage the actual number of tourists is low therefore requiring little management intervention. However, an increasing awareness of the site through publicity, even media hype, or word of mouth results in increased numbers of visitors leading to a greater demand for facilities and increased pressure on ecological and social conditions. This latter kind of tourist, who increasingly patronises the site, can be classified as generalist. They are often thought to have little special interest in the site's wildlife and rely heavily on the development of supportive infrastructure. The outcome typically results in impacts on the wildlife and their environment and additional management intervention is subsequently needed (Duffus and Dearden 1990; Higham 1998).

With the site/wildlife tourism experience becoming more developed to cater more for the generalist wildlife viewer, it will often no longer attract the specialist wildlife viewer. Instead, specialists are displaced to other less developed and more remote sites. This is known as site user evolution (Duffus and Dearden 1990). Increasing visitor numbers and the development of facilities to cater for the more generalist visitor is more likely to have an impact on the wildlife species and the natural environment. Greater management of the site is therefore needed in order to accommodate higher visitation and ameliorate negative environmental and social impacts. When this type of change takes place, the management response is often to develop the site to rectify an impact situation or address increasing visitor numbers. This may result in increasing levels of impact, further (new) impacts, cumulative impacts and, as mentioned previously, displacement of specific user types. In response to these changes there is usually a greater need for management of the site in order to accommodate higher visitation and ameliorate negative impacts. The situation does not usually include management actions such as limiting use or site restoration such as closing or resting a site. It does, however, usually include an increase in visitor communication and education (Smith and Newsome 2006).

Accordingly, this chapter explores the question of whether a specialist or generalist approach is better for the long-term sustainability and management of last chance wildlife tourism interactions. Different destinations and corresponding management agencies are taking differing approaches. Managers of protected areas, where much wildlife tourism takes place, face the dilemma of promoting wildlife tourism to earn revenue and conservation of the wildlife and their habitat

(Goodwin and Leader-Williams 2000). Balancing wildlife tourism and conservation is becoming more difficult for managers as an increasing number of tourists explore more remote places (Goodwin and Leader-Williams 2000).

Through the use of case studies focusing on rare primates and birds, the complexities of presenting and managing the concept of 'last chance wildlife tourism' are now considered further. The case studies illustrate current examples of existing tourism operations which could be deemed specialist and/or generalist wildlife tourism according to the concept as originally presented by Duffus and Dearden (1990).

## The case of rare primates

Larger primates and small monkeys with a limited geographical distribution are in decline globally. Examples include the hoolock gibbon (*Hylobates hoolock*) in northeast India, the black gibbon (*Nomascus concolor*) in China and Laos and the proboscis monkey (*Nasalis lavatus*) in Borneo. In Africa the bonobo (*Pan paniscus*) may be the first great ape to become extinct in the wild (Garbutt and Unwin 2007). Also in Africa population declines due to habitat loss and hunting are evident for species such as mandrill (*Papio sphinx*) and drill (*Papio leucophaeus*). In South America the uakari (*Cacojao calvus*) and various species of marmoset such as the golden lion tamarin (*Leontopithecus rosalina*) are in decline and are species that either are, or could be, of tourism potential as part of a wider forest ecotourism experience. In Madagascar at least eight species of lemur are under significant threat, while at the same time being the subject of nature-based tourism activities. An example of this is the indri (*Indri indri*) in Perinet Reserve where two groups are habituated to tourist visitation. In all cases the causes of decline include habitat destruction, climate change, increasing demands for agricultural land, logging, forest fragmentation, conversion of natural vegetation to plantations, poaching for the illegal pet trade, hunting for bush meat, traditional medicines and in some cases political instability and war.

Two examples of rare primate tourism are considered here: (1) the orang-utan (*Pongo sp*) in southeast Asia and (2) the mountain gorilla (*Gorilla beringei beringei*) in central Africa. In the case of the orang-utan, there are two species: Sumatran orang-utan (*Pongo abelii*) endemic to Sumatra and the Bornean orang-utan (*Pongo pygmaeus*) occurring in Borneo. Both species are considered to be at risk of extinction in the wild within the next 10–20 years (Ferber 2000; Nellerman *et al.* 2007). Tourism issues surrounding orang-utan tourism in north Sumatra have previously been considered by (Newsome *et al.* 2002), so attention is paid to tourism surrounding rehabilitated and wild orang-utans in Sabah.

## Tourism at the Sepilok orang-utan rehabilitation centre – a generalist attraction

Tourists who visit Sepilok are virtually guaranteed orang-utan sightings, as it is a rehabilitation centre. The centre itself has been in existence since 1964 and the centrepiece of a programme to cater for displaced, orphaned and injured orang-utans, in this case as a result of deforestation, conversion of natural habitat to oil palm plantations and due to the taking of young animals for the illegal pet trade. Orang-utans undergo a rehabilitation process that involves captive animals undergoing medical examination and a period of quarantine to prevent any disease spreading to the existing rehabilitated population. Orphaned juveniles need to be assisted in terms of forest survival and are gradually moved to more natural settings where human support is gradually decreased so that the animals are able to cater for themselves with minimal and eventually no assistance at all. An important component of the rehabilitation process is supplemental feeding with milk and fruits such as bananas. Fully rehabilitated orang-utans are released into the Kabili-Sepilok Forest Reserve (4,300 ha) with the aim that they become fully integrated with an already rehabilitated and wild population of orang-utans.

This wildlife tourism interaction can be classed as a generalist tourism experience and involves observing the orang-utans being fed at a public observation deck (Figure 5.3). Orang-utans are called to the deck at 10 am and 3 pm daily and there are rules in place and rangers are present to ensure visitor compliance with guidelines such as no contact or feeding the animals, maintenance of security over carried items that can be taken by orang-utans and restrictions on the carrying and use of insect repellents which may prove attractive to the animals. Infrastructure development at the rehabilitation centre to facilitate access and educational content comprises a visitor information centre and boardwalk that leads to the feeding

*Figure 5.3* Sepilok Orang-utan Rehabilitation Centre. Orang-utans are called to a feeding platform from which they can be observed by visitors, who may arrive independently, on private tours or as part of a coach tour (Photo credit: David Newsome)

platform. Visitors may also undertake forest walks on marked trails, where sightings of rehabilitated orang-utans are possible. Visitors also have the opportunity to visit a wildlife observation hide.

Clearly the experience is not of a fully wild orang-utan but Sepilok provides the scope for tourists to see these animals in a natural setting rather easily. Easy access coupled with the content provided by some guided tour commentaries at the feeding sessions at least provides an opportunity for the generalist wildlife tourist, who might also be termed the 'soft ecotourist', who has more casual nature-related interests (Fennell 2008), to learn about and experience a rare and endangered species. However, within the context of the generalist wildlife visitor viewpoint presented here, it is more likely that the orang-utans are dominantly a photographic collectable, an attitude that is more likely to be at odds with the view expressed by Hilary Brandt earlier in this chapter (Russell and Ankenman 1996). Furthermore, the visitor motivation to see and photograph orang-utans may be a result of their physical attributes and human-like qualities (Tremblay 2002) rather than a concern to care about them and their conservation, an attitude that can be interpreted to be in line with the generalist perspective that is being argued here.

## Viewing orang-utans along the Kinabatangan river and ecotourism at Sukau Rainforest Lodge, Sabah – a specialist attraction

The Kinabatangan River floodplain provides the opportunity to see wild orang-utan, and in contrast to the Sepilok experience would be a very difficult objective to achieve without tour operator support. A vigorous nature-based tourism industry has grown along the banks of the Kinabatangan River with most operators providing guided boat trips and forest walks from which a range of wildlife can be seen. Wildlife resources include many typical Asian rainforest species but of particular interest, and desired visitor sightings, are Bornean pygmy elephant (*Elephas maximus borneensis*), bearded pig (*Sus barbatus*), four species of monkey, orang-utan, estuarine crocodile (*Crocodylus porosus*) and many birds such as storms stork (*Ciconia stormi*), eight species of hornbill, bat hawk (*Macheiramphus alcinus*) and buffy fish owl (*Ketupa ketupu*). Although visitors visit the Kinbatangan floodplain for a total rainforest experience, of particular significance is the opportunity to sight Bornean orang-utans in the wild. This wildlife tourism experience attracts more specialist wildlife tourists.

The proliferation of tourist facilities over the last 15 years has resulted in a range of wildlife experiences being offered in terms of customer service, commitment to sustainability and the quality of environmental education. Most tourism operators recognise the economic opportunities that the floodplain region provides as an alternative to the rapid expansion of oil palm plantations that has occurred throughout Borneo during the last 20 years.

A tour operation which aims to extend the scope of wildlife viewing to rainforest appreciation and the prevention of habitat loss for orang-utans and other species falls under the umbrella of Borneo Ecotours, a company which own the Sukau Rainforest Lodge. The lodge is located on a private land purchase on the banks of Kinabatangan River and is a two hour boat ride or 130 km overland from the town of Sandakan in east Sabah. Ecotourism at the Lodge consists of authentic natural experiences, wildlife sightings, facilities (boardwalks, boat trips) to enhance the experience, opportunities for clients to learn, committed management and a comprehensive programme of stakeholder engagement and participation. In terms of design, the lodge is built on stilts and situated 30 m from the riverbank according to a policy of native vegetation retention. There is a philosophy of water (rainwater tank supply) and energy (noise-insulated generator and solar supplied) conservation. Solid waste is transported out and septic tanks have been installed to deal with toilet wastes. Recycling is in place and strongly encouraged.

Tourism operations and activities comprise guided boat tours on the main river and its tributaries and associated ox-bow lakes. Walk trails have been developed at the lodge (boardwalk) and leading off from a landing point at an ox-bow lake. There are seven permanently employed guides who are trained in-house (three-week course) in wildlife spotting and interpretation and boat handling. Sightings of orang-utan in particular are not guaranteed and it is explained to visitors that the river floodplain is a wild habitat and a different experience to viewing orang-utans at Sepilok.

Most wildlife viewing takes place from a boat and the viewing experience becomes more personal as a single boat trip is limited to eight persons. More than one boat may, however, depart from Sukau at a time but different tributaries and river locations are used to view wildlife. During wildlife observation electric motors and oars are used to minimise noise and exhaust from engines and guides identify wildlife and offer interpretation of observed wildlife.

Sukau Lodge management is committed to sustainability and conservation and provides opportunities for tourists to become involved. There is evidence of the deeper engagement according to the perspectives raised by Hilary Brandt (Garbutt and Unwin 2007). Evening slide shows that are delivered usually to an audience of around 10–15 guests contain strong conservation messages. Visitors are also encouraged and provided with the opportunity to plant trees on the edge of oil palm plantations, along the riverbank and in previously logged areas, and the lodge has an active programme of local community engagement. This clearly is not a mass tourism, generalist experience but at the same time is not entirely focused on one particular species. It does, however, provide the scope for seeing orang-utan in the wild. It could even be argued that it is a more exlusive specialist experience by virtue of relative expense but it cannot be denied there is a real attempt to engage the visitor in a deeper, more personal experience of the Bornean rainforest that has an ultimate conservation objective.

Having highlighted this as a possible example of specialist interest in rare primate tourism, there are, however, problems. There are environmental issues emanating from the wider landscape in terms of erosion and pesticide/fertiliser issues associated with the oil palm industry. In addition to this, there is the problem of uncontrolled tourism development in the form of lodge proliferation and excessive numbers of boats. Noise, fumes and crowding haves been reported at wildlife viewings and the behaviour of some tour operators is problematic in terms of the quality of guiding and speeding along the river in order to ensure clients gain a view of wildlife reported by other operators. Such activity could result in reduced visitor satisfaction and a decline in repeat visitation. It also raises the question as to whether unplanned and inappropriate tourism development could impact on operators fostering a specialist wildlife tourism/rainforest experience and the creation of a shift to the issues associated with more generalist wildlife tourism.

Sukau Lodge already makes a significant contribution towards a specialist wildlife tourism profile by promoting environmental awareness, environmental waste disposal, engaging in recycling and promoting the sustainable use of energy. But do all the lodges that occur along the lower Kinbatangan floodplain do this? More importantly will new operators do this? Tourism operators who have to 'cut corners' in the face of intense competition may end up offering reduced quality ecotourism experiences and have less time for the promotion of more time-consuming and potentially expensive sustainable tourism and conservation.

## Mountain gorilla tourism – specialist and generalist attraction

Mountain gorilla (*Gorilla beringei beringei*) tourism has been operating since the mid- to late 1970s (Briggs and Booth 2009) with tours taking place in Uganda, Rwanda and the Deomocratic Republic of Congo (DRC) in Africa. There are two distinct populations of mountain gorillas. The first is found in Bwindi Impenetrable National Park in Uganda while the second is found in the Virunga Volcano Region, which covers the international borders of Rwanda, Uganda and the DRC. The mountain gorilla is currently listed as Endangered by the International Union for the Conservation of Nature and Natural Resources (IUCN) (Robbins and Williamson 2008).

Mountain gorilla tourism is big business with costs of $US500 per day per person to view the gorillas (Garbutt and Unwin 2007; Carwardine 2009). The official permits are administered by the national parks. There are numerous tour companies that offer the chance to interact with the gorillas, from the very budget overland camping safaris through to the more luxurious and expensive. It is also possible to organise your own tour. Many of the overland adventure companies appeal to the younger market, offering the opportunity to view the mountain gorillas as a part of their long haul overland treks across the African continent.

It is due to the many different tour options for mountain gorilla interactions that makes it difficult to classify this tourism activity as either a specialist or generalist attraction. In many ways it can easily be thought of as a specialist attraction as it is difficult to travel to, is expensive, and a permit limits numbers. However, due to the popularity of the overland tour many of the tourists who are coming to view the gorillas would not be thought of as 'specialist' wildlife tourists. There are increasing number of visitors who are often not experienced or with great knowledge (Higham *et al.* 2008) as many are coming to view the gorillas because it is a part of their long haul 'adventure tour' rather than a special desire to see the wildlife. Many of the younger tourists are there to view the gorillas as it is often seen as an added bonus rather than a special desire. In addition, due to increased visitor numbers tourism-related facilities have become developed and diverse (Sandbrook 2010). Interestingly, for mountain gorilla tourism there has been little site user evolution due to the limited number of places to interact with the mountain gorillas, therefore the specialist wildlife viewer will continue to visit.

Mountain gorilla tourism, as a specialist and/or generalist attraction, has become a universally accepted tool to foster the conservation of the gorillas. It has made a significant contribution to conservation efforts of gorilla populations. The International Gorilla Conservation Program noted that the income from tourism for the gorillas has been probably the most important factor in contributing to the support of parks and conservation of habitat over the decades (Kirby 2002). As Klailova, Hodgkinson, and Lee (2010: 897) note: 'In theory, ecotourism should represent a win–win situation for both the country in which it is held and for the vulnerable places and animals that it protects.' However, this is not always the case.

Risks of disease transmission between gorillas and humans are associated with habituation and tourist visitation. Gorillas are susceptible to many human diseases with potentially devastating consequences because of their genetic relatedness with humans. Gorilla tourism rules have been established to minimise health risks to the wildlife with rules on animal viewing distances (more than 7 metres), visit duration time (less than one hour), tourist group size (less than ten) and number of times the group of gorillas can be visited each day (preferably one group each day) (Butynski and Kalina 1998; Klailova *et al.* 2010; Muehlenbein and Ancrenaz 2009; Newsome *et al.* 2005). However, in many cases these are not always abided by with tourists having closer interactions than allowed. There is no doubt that tourism is contributing to the conservation of the mountain gorillas. Yet, care is needed to continue to find the right balance. Whether this is achieved by following a more specialist approach to mountain gorilla tourism or encouraging the more generalist type of tourist remains an uncertainty.

## The case of rare birds

Birdlife International (2000) has listed many rare and endangered species of birds including species such as the Victoria crowned pigeon (*Goura victoria*), orange bellied parrot (*Neophema chrysogaster*), hyacinth macaw (*Anodorhynchus hyacinthinus*), Philippines eagle (*Pithecophaga jefferyi*), gouldian finch (*Erythrura gouldiae*), little spotted Kiwi (*Apteryx owenii*), Bali starling (*Leucospar rothschildi*) and 16 species of albatross (e.g. *Diomedia sandfordi*). The causes of decline (some sources indicate that 70 per cent of the world's birds are declining, WWF, 2009) are varied depending on location, habitat and the regional impacts of climate change (Butchant *et al.* 2010). Forest clearance and habitat modification are major drivers in terrestrial ecosystems as considered previously. Seabirds, such as albatrosses, are subject to the impacts of long-line fishing and the effects of introduced species at island breeding locations. The increased pollution of the marine environment, especially in the form of plastic flotsam, is also a problem as plastic wastes are often eaten and regurgitated to chicks. This is thought to be a significant cause in the decline of albatross species.

Birds form the centre-piece of a very popular natural history activity and hobby that spans all ages and nationalities. Bird watchers form the largest percentage (millions of participants) of wildlife watchers globally (Moss, 2009) (for additional information on the birding industry see chapter 6 this volume). In addition to this, bird watchers are a very complex wildlife watching demographic with casual, mildly interested bird watchers at one end of the spectrum to 'hard core twitchers' at the other who go to great expense and personal inconvenience to record the sighting of a species they have not seen before. Some tourism providers set themselves up as bird watching destinations as private lodges or as accommodation facilities in the vicinity of suitable protected areas. Bird watching can be carried out casually, be the prime activity of free-independent travellers, take the form of specialised tours and wildlife watching packages and some sites may be a stopover point for coach tours. Most categories of bird watcher have an interest in seeing rare and endangered birds. While many birds can, and are, the focus of tourism interest, Carwardine (2009) is of the view that carefully managed specialist tourism may contribute to a wider conservation effort and the protection of rare birds.

Examples of specialist bird watching tourism are thus evident globally and bird watchers are catered for by bird watching tour groups (e.g. BirdQuest and Ornitholidays) that focus on a plethora on destinations worldwide. One example of such a destination is New Zealand where little spotted Kiwi (*Apteryx owenii*), classed as vulnerable because of its small range and susceptibility to introduced predators, was introduced to Kapati Island around early 1900s and released on Tiritiri Matangi in 1993 and 1995 (Birdlife International, 2000). Small captive breeding populations are kept for public education purposes and bird watching tourists can visit Kapati and Tiritiri Matanga according to a visitor permit system. Both islands also provide the visitor with the opportunity to see other rare species such as takahe (*Porphyrio hochstetteri*), kaka (*Nestor meridionalis*) and saddleback (*Philesturnus carunculatus*) on Kapati and takahe, whitehead (*Mohoua albicilla*), North Island robin (*Petroica longipes*), stitchbird (*Notiomystis cinta*) and saddleback on Tiritiri Matangi (for a discussion on the kakapo see Lemieux and Eagles this volume). Two examples are briefly described here: the first being the Bali starling of Indonesia, which requires access by permit and specialist bird guiding services; and the Royal albatross colony at Taiaroa Head in New Zealand, a specialist bird watching destination which evolved according to the Duffus and Dearden model into a generalist wildlife tourism destination.

The Bali starling (endemic to Bali) is one of the world's rarest birds and occupies a small range (ca. 40 $km^2$) in Bali Barat National Park. The bird is critically endangered primarily as a consequence of illegal capture for the cage bird trade. Although protected under Indonesian Law since 1979, the birds continued to decline to a population of 12 at around 2000 (Birdlife International 2000). Park mismanagement and corruption was attributed to this when in the mid-1990s an armed gang took some 39 birds that were awaiting release into the wild from the Bali Starling Conservation Centre.

The Bali starling has always been a focus for specialist birdwatchers and over time bird watching tourism has continued to develop in Bali with several specialist guides now offering regular bird guiding services. All these tour guides emphasise the Bali starling and handle the associated permit and logistics of visiting the Bali starling project site in the park, which offers the best possibility of seeing a starling in the wild. Clients are taken to the Teluk Telor ranger station on the Prapat Agung Peninsula. Wild and captive-bred released birds can be seen in the vicinity of a large captive enclosure. Given that Bali is a mass tourism destination, it is very notable that very few visitors, other than specialist bird watchers, would take the time and spend the money on what some would describe as a unique bird watching experience.

In the case of albatross viewing at Taiaroa Head in New Zealand, Higham (1998) describes a situation where in the 1970s only about 30 people a week would visit a breeding colony of northern albatross. Over time the site, and particularly the rare birds, were promoted and marketed as a visitor attraction and this, in combination with increased tourism in New Zealand, generally resulted in an increase in people visiting the site. Over time, as Taiaroa Head became part of coach tour group scheduling, a different type of visitor to that of the early days of visitation was delivered to the site. In conjunction with this, the increased visitation, from around 300–700 a year in the 1970s to 40,000 a year by 1990, stimulated a programme of improved access, expanded car parking and visitor facility development. Specialist bird watchers were subsequently displaced by a more complex demographic, which included individuals who required policing and supervision. The increased visitation and disturbance resulted in negative impacts on the birds (see Higham 1998), which is counter to the principles of sustainable tourism as desired by Carwardine (2009), conservation biologists and protected area managers alike.

## Conclusion

The impact of humans on wildlife and natural areas is extensive and continues on despite a plethora of conservation agencies and expanded protected areas around the world. Major problems include land clearing and pollution. Nature-based tourism is one of a range of mechanisms that has been, and can be, put into place to engage humans with the value of nature and wildlife. Accordingly managers of protected areas, where much wildlife tourism takes place, face the dilemma of promoting wildlife tourism to earn revenue and conservation of wildlife and their habitat (Goodwin and Leader-Williams 2000). Through the use of a number of case studies focusing on rare primates and birds, the complexities of managing last chance wildlife tourism can be considered and questioned.

In a re-appraisal of Duffus and Dearden's (1990) wildlife tourism framework Catlin *et al.* (2011) reiterate that generalists sit at the opposite end of the wildlife tourism spectrum from that of specialists, a situation that necessitates additional facility development and more mediation between themselves and the target species. Such conditions have the potential to place additional stress on the wildlife and surrounding environment. Specialists, on the other hand, have a deeper interest in the target species, require knowledge and are more likely to be interested in conservation. It could thus be argued that a specialist approach to last chance wildlife tourism is best for fostering human interest, care and the long-term conservation of wildlife. This in itself may not be a straightforward argument, as Dearden *et al.* (2006) note that previously specialised tourism activities are now being undertaken by a wider tourism demographic which potentially blurs the distinction between specialist and generalist wildlife viewers.

For rare primates, is the generalist approach taken at Sepilok Orang-utan Rehabilitation Centre better for the conservation of the wildlife than the specialist viewing of the orang-utans along the Kinabatangan River? Or is there a need for both to ensure the long-term sustainability of the orang-utan. In the case of the Bali starling, would it have a greater chance of survival if its plight was better advertised and promoted in a tourism context, therefore potentially attracting a more generalist type of tourist? Such an approach could bring in more money for conservation but what would be the impact on the starling and its habitat?

Clearly such a trend from the specialist to the general visitor in the context of Duffus and Dearden's model needs to be evaluated with caution. Figure 5.1 illustrates a group of visitors on a tiger-viewing safari in India where most of the people are likely to be generalists. The reasoning behind such an assertion is that zoological gardens, television, radio, magazines, newspapers, Youtube and websites have profiled the plight of the tiger as a vanishing species. Such profiling has raised awareness and expanded the demographic now interested is seeing this species in the wild. This tourism demographic is now likely to consist of a mix of generalists and even some specialists who are taking their opportunity to see a rare species where it can be reliably viewed. Last chance tourism can then be a mix of generalist and specialist clients. It may even be that some last chance tourists are not wildlife tourists at all and they simply go on a tour to see and photograph something that is special in a temporal context. Last chance tourism may then vary according to the species concerned and degree of difficulty in locating and viewing a species. Some situations therefore may be mostly generalist in nature while others – for example, in the case of rare birds – almost exclusively specialist in nature.

Regardless of the situation, increased visitation, development and revenue has the potential to further profile the wildlife to a wider audience, raise awareness and provide money for conservation. On the other hand, though misguided tourism development, the presence of large numbers of people and the need to manage crowds and certain individuals needs forward planning, adaptive management, supervision and appropriate site interpretation along with conservation related education. Increasing our understanding and engaging in further dialogue about the nature of last chance tourism is very important in setting the agenda for future research and managing the nature of this tourism phenomenon.

## References

Adams, D., and Carwardine, M. (1990) *Last Chance to See*. New York: Ballantine Books.

Ballantyne, R., Packer, J. and Hughes, K. (2009) 'Tourists' support for conservation messages and sustainable management practices in wildlife tourism experiences', *Tourism Management* 30: 658–664.

Birdlife International (2000) *Threatened birds of the world*. Barcelona and Cambridge: Lynx Edicions and Birdlife International.

Briggs, P., and Booth, J. (2009) *Rwanda, Bradt Travel Guides Ltd*. Bucks, UK: Bradt Tour Guides.

Bryan, H. (1977). Leisure value systems and recreation specialization: The case of trout fishermen. *J. Leisure Res.* 9: 174–87.

Bryan, H. (1980). 'Sociological and psychological approaches to assessing and categorizing wildlife values' in *Wildlife Values*, ed. W.W. Shaw and E.H. Zube. *Center for Assessment of Noncommodity Natural Resource Values, Inst. Ser. Rep.* 70–6.

Butchant, S., Collar, N., Statterfield, A. and Bennun, L. (2010) 'Conservation of the world's birds: the view from 2010', in del Hoyo, J., Elliott, A. and Christie, D. A. (eds.) *Handbook of the Birds of the World*. Vol. 15: Weavers to New World Warblers. Barcelona: Lynx Edicions.

Butler, R.W. (1980). The concept of a tourist area cycle of evolution: Implications for management of resources. *Can. Geogr.* 24: 5–12.

Butynski, T., and Kalina, J. (1998) 'Gorilla Tourism: A critical look', in E.J. Milner-Gulland and R. Mace (eds.) *Conservation of Biological Resources*. Oxford: Blackwell Science.

Catlin, J., Jones, R. and Jones, T. (2011) 'Revisiting Duffus and Dearden's wildlife tourism framework', *Biological Conservation*. 144: 1537–44.

Carwadine, M. (2009) *Last chance to see: in the footsteps of Douglas Adams*. London: Collins.

Courchamp, F., Angulo, E., Rivalan, P. Hall, R.J., Signoret, L., Bull, L. and Meinard, Y. (2006) 'Rarity value and species extinction: the anthropogenic allee effect', *PLoS Biology* 3(12): 2405–2410.

Curtin, S. (2009) From 'Sandals and Beards to Sophisticated Urbanites'. In *CAUTHE – 18th International Research Conference: See Change: Tourism and Hospitality in a Dynamic World*. Fremantle, WA.

Dearden, P., Bennett, M. and Rollins, R. (2006) 'Implications for reef conservation of diver specialization', *Environmental Conservation* 33: 353–363.

Duffus, D.A., and Dearden, P. (1990) 'Non-consumptive wildlife-orientated recreation: a conceptual framework', *Biological Conservation* 53: 213–231.

Easier Inc. (2011) 'Tour operator launches "Last Chance To See" itineraries'. Online: www.easier.com/55858-tour-operator-launches-last-chance-to-see-itineraries.html (accessed 5 April 2011).

Fennell, D. (2008) *Ecotourism*. London and New York: Routledge.

Ferber, D. (2000) 'Orangutans face extinction in the wild', *Science* 288: 1147–1149.

Garbutt, N., and Unwin, M. (2007) *100 Animals to see before they die*. Bradt Travel Guides, Globe Pequot Press, USA.

Goodwin, H.J., and Leader-Williams, N. (2000) 'Tourism and protected areas – distorting conservation priorities towards charasmatic megafauna?' in A. Entwistle and N. Dunstone (eds.) *Priorities for the conservation of mammalian diversity*. Cambridge: Cambridge University Press.

Hall, R.J., Milner-Gulland, E.J. and Courchamp, F. (2008) 'Endangering the endangered: the effects of perceived rarity on species exploitation', *Conservation Letters* 1: 75–81.

Higham, J.E.S. (1998) 'Tourists and albatrosses: the dynamics of tourism at the Northern Royal Albatross Colony, Taiaroa Head, New Zealand', *Tourism Management* 19: 521–531.

Higham, J.E.S., Lusseau, D. and Hendry, W. (2008) 'Wildlife viewing: The significance of the viewing platforms', *Journal of Ecotourism* 7 (2): 137–146.

Kachel, U., and Jennings, G. (2010) 'Exploring Tourists' Environmental Learning,

Values and Travel Experiences in relation to Climate Change: A postmodern constructivist research agenda', *Tourism and Hospitality Research* 10(2): 130–140.

Kirby, A. (2002). Hope Rise for Mountain Gorillas. *BBC News World Edition.*

Klailova, M., Hodgkinson, C. and Lee, P.C. (2010) 'Behavioural responses of one western lowland gorilla (*Gorilla gorilla gorilla)* group at Bai Hokou, Central African Republic, to Tourists, Researchers and Trackers', *American Journal of Primatology* 72(10): 897–906.

Lemelin, H., Dawson, J., Stewart, E.J., Maher, P. and Lueck, M. (2010).'Last-chance tourism: the boom, doom, and gloom of visiting vanishing destinations', *Current Issues in Tourism* 13(5): 477–493.

Moss, S. (2009) 'Birding past, present and future-a global view', in del Hoyo, J., Elliott, A. and Christie, D.A. (eds.) *Handbook of the birds of the world.* Vol. 14: Bush-shrikes to Old World Sparrows. Barcelona: Lynx Edicions.

Muehlenbein, M.P., and Ancrenaz, M. (2009) 'Minimising pathogen transmission at Primate Ecotourism Destinations: The need for input from travel Medicine', *Journal of Travel Medicine* 16(4): 229–232.

Nellerman, C., Miles, L., Kaltenborn, B.P., Viture, M. and Ahlenius, H. (2007) *The last stand of the Orangutan: state of emergency: illegal logging, fire and palm oil in Indonesia's National parks.* Arendal, Norway: UNEP.

Newsome, D., Moore, S.A. and Dowling, R.K. (2002) *Natural Area Tourism.* Clevedon, UK, and Buffalo, NY: Channel View Publications.

Newsome, D., Dowling, R. K. and S. A. Moore. S. A. (2005) *Wildlife Tourism.* Clevedon, UK, and Buffalo, NY: Channel View Publications.

Orangutanadventures.com (no date) Orangutan Tour Central Kalimantan, Borneo. Online: www.orangutanadventures.com/ (accessed 5 April 2011).

Reynolds, P.C. and Braithwaite, D. (2001) 'Towards a conceptual framework for wildlife tourism' *Tourism Management* 22: 31–42.

Robbins, M., and Williamson, L. (2008) '*Gorilla beringei*'. *IUCN Red List of Threatened Species*, edited by IUCN. Version 2010, online: www.iucnredlist.org/ (accessed 15 March 2011).

Rodger, K., Moore, S.A. and Newsome, D. (2007) 'Wildlife tourism in Australia: Characteristics, the place of science and sustainable futures', *Journal of Ecotourism* 15(2): 160–179.

Russell, C. and Ankemanm, M.J. (1996) 'Orangutans as photographic collectibles: Ecotourism and the commodification of nature', *Tourism Recreation Research* 21: 71–78.

Sandbrook, C.G. (2010). 'Local economic impact of different forms of nature-based tourism', *Conservation Letters* 3(21–28).

Shackley, M. (1996) *Wildlife tourism.* London: International Thomson Business Press.

Smith, A.J. and Newsome, D. (2006) *An investigation into the concept of and factors leading to impact creep and its management.* The Gold Coast, Queensland: Cooperative Research Centre for Sustainable Tourism.

Tremblay, P. (2002) 'Tourism wildlife icons: attractions or marketing symbols', *Journal of Hospitality and Tourism Management* 9: 164–180.

WTO (2003) 'Climate change and tourism'. Paper read at Proceedings of the 1st international conference on climate change and tourism, 9–11 April, Djerba, Tunisia.

WWF (2009) 'Where have all the birds gone?'. Online: www.brighthub.com/environment/science environmental/articles/12045.aspx (accessed 21 March 2011).

# 6 Last chance birding

## Twitching to see it first or last?

*Glen T. Hvenegaard*

## Introduction

Last chance tourism is a "niche tourism market where tourists explicitly seek vanishing landscapes or seascapes, and/or disappearing natural and/or social heritage" (Lemelin *et al.* 2010: 478). Many people will want to see such attractions today if there is a chance they will be irreversibly gone in the future. In order to understand, accommodate, and manage the interests and behaviors associated with last chance tourism, one must explore a complex set of interactions among the provision of information, offering of tourism products, personal decision making, and management of the tourism resources. Last chance tourism has been analyzed for tourism opportunities threatened by climate change (Becken and Hay 2007) and many other forces (Addison 2008; Hughes 2008). Analyses are needed for other tourism opportunities facing similar declines.

One tourism sector facing declines in its target resource is bird watching tourism. Bird extinctions are occurring at a rate 100 times faster today than the known background rate before human impacts (Pimm *et al.* 2006). The continued existence of about 12 percent of all bird species is threatened by a variety of human impacts, especially habitat change (BirdLife International 2008; Lebbin *et al.* 2010). Moreover, regardless of their status as endangered or threatened, the populations of many bird species are in decline (BirdLife International 2008). With many species facing imminent extinction, and predictions for more declines or extinctions in the future, bird watchers may plan trips to take advantage of a "last chance" to see certain bird species. The popularity of bird watching has grown rapidly in the last few decades throughout the world, but particularly in developed countries. For example, in 2006, 48 million Americans participated in birding activities (21 percent of the population), spending about US$36 billion on trip-related expenditures and equipment (Carver 2009).

People want to watch birds for a variety of reasons. Some people watch birds to appreciate characteristics of the birds, while others watch birds to socialize with friends, improve identification skills, promote bird conservation, collect research data, or compete with fellow bird watchers (Eubanks *et al.* 2004; McFarlane 1994; Scott *et al.* 1999). People who watch birds can be categorized in many ways. Oddie (1995) recognizes five groups (ornithologist, birdspotter/birdfancier, birdwatcher, birder, and twitcher), of which the latter three are most relevant to this chapter

*Figure 6.1* Birders at the Fraser Valley Bald Eagle Festival set up scopes to view Bald Eagles near the confluence of the Chehalis and Fraser Rivers, British Columbia

on last chance birding. *Birdwatchers* casually watch and identify birds as part of their leisure time pursuit (Schaffner 2009). *Birders* actively pursue birds, and "not merely looking with pleasure, but are interested in scientific classification, environmental issues, and the more detailed study of birds" (Connell 2009: 204). According to Sheard (1999), birders study, identify, and collect birds. *Twitchers* approach birding with extra competitive intensity, as shown by their strong desire to see rare birds, emotional involvement in the sport, and willingness to readily travel to observe new birds (Cocker 2001; Connell 2009; Oddie 1995; Sheard 1999). Birders, and especially twitchers, maintain bird lists which usually include species identified in a given geographic area and time period (Schaffner 2009). Many birders, and most twitchers, keep track of a "life list", a record of all bird species identified in a lifetime. "Birder" is largely an American term, while "twitcher" is largely British. This chapter will use "birder" and "birding" throughout unless the research or examples apply specifically to "twitchers" or "twitching".

Birders have also been differentiated with reference to their level of specialization, which is a "continuum of behaviour from the general to the particular, reflected by equipment and skills used in the sport and activity setting preferences" (Bryan 1977: 175). Typically, more specialized (or advanced) birders have more specific motivations, are more experienced, own and use more birding equipment, spend more time birding, and feel that birding is more central to their lifestyle than less specialized (or beginner or novice) birders (Hvenegaard 2002). This chapter will refer to "birders" most often, but will insert specialization terms where appropriate.

*Figure 6.2* Birders and bird photographers used rare bird alerts to find the rare Kirtland's Warbler at Point Pelee National Park, Ontario

Most people view birds largely on their own, but occasionally as part of organized tours, which can be commercial or non-commercial (see Figure 6.2). Birders and twitchers are interested in all birds, but have special interest in rare or unusual species because these birds are valuable in terms of novelty and potential for adding to a bird list. This focus on rare birds connects bird watching tourism to the larger sector of "last chance tourism", in that birders want to view species before they become extinct or before their populations decline to levels which make their viewing no longer feasible. The goal of this chapter is to explore the concept of rarity in birds, why birders are interested in rare birds, possible explanations for that interest, and issues in managing birding tourism that targets rare birds.

## Background on rare birds

An overview of rarity among bird populations is required to provide a basis for human reactions in the form of birding tourism. Birds are a very well-known group of species (Pimm *et al.* 2006). Nearly 10,000 bird species have been described worldwide (Birdlife International 2008), with new species being added regularly through discovery and taxonomic reassessment (Newton 2003). Bird species diversity is highest in the Neotropical geographic realm (3,370 species), followed by the Afrotropical (1,950), Indomalayan (1,697), Australasian (1,592), Palearctic (937), Nearctic (732), and Oceanic (187) realms (Newton 2003). Many bird species exist in very small ranges (BirdLife International 2008). In fact, over 35 percent of all bird species are endemic to one of 34 hotspots for biodiversity identified by Conservation International (2011). Within these hotspots, only 14 percent of the habitat remains in a natural condition (Conservation International 2011).

According to Pimm *et al.*'s (2006) comprehensive review, since 1500, about 1.3 percent of the known bird species have become extinct. This translates into a rate of about 26 extinctions per million species per year (E/MSY), which is much higher than the background rate of 1 E/MSY before human impacts. However, this rate is likely as high as about 100 E/MSY if one accounts for the timing and rates of bird extinctions and discoveries and of the process of declaring a species extinct. At present, 1,226 bird species (about 12 percent) are threatened due to small and declining populations and ranges (BirdLife International 2008). Of these, 55 percent are vulnerable, 30 percent are endangered, 15 percent are critically endangered, and less than 1 percent are extinct in the wild (BirdLife International 2008). Pimm *et al.* (2006) estimate that the rates of endangerment and extinction will increase throughout the twenty-first century, due to a loss of low population species, declaration of already extinct species, and greater intensity of critical threats. The percentage of threatened species is especially high for large species with low reproductive rates (e.g. 80 percent for albatrosses, 60 percent for cranes, 27 percent for parrots, and 23 percent for pheasants; BirdLife International 2008). Even once common species, such as the Western Meadowlark (*Sturnella neglecta*) and Horned Lark (*Eremophila alpestris*) from North America's grasslands, are declining, especially in the last 20 years (BirdLife International 2008; Brennan and Kuvlesky 2005; McCracken 2005).

The major threat to bird populations is habitat conversion, fragmentation, and degradation (Lebbin *et al.* 2010). Habitat change results from development, mining, agriculture, grazing, forestry, and a variety of other human uses of the land. Other major threats to bird populations include invasive and overabundant species (e.g. through predation, deliberate species introduction, competition, and disease), collisions with human-made structures, exploitation, bycatch from fishing, pollution, and climate change (BirdLife International 2008; Lebbin *et al.* 2010; Price 2003). Climate change, in particular, will prompt habitats to shift into new areas, but existing human uses may limit the effectiveness of that shift by birds and their habitats. These threats have disproportionately impacted habitats such as oceanic islands, forests, wetlands, coastal areas, grasslands, shrublands, and riparian areas (Brennan and Kuvlesky 2005; MacDonald 2003; McCracken 2005; Sekercioglu *et al.* 2004). These threatened habitats have most of the world's threatened species.

How will these trends impact the decisions of birders wanting to see rare species? This depends on birders' concepts of, and values associated with, rarity. In addition, true rarity may differ from perceived rarity (Hall *et al.* 2008). For example, a species perceived to be rare may actually be "widespread but at low densities throughout its range, . . . found in only a few locations, . . . locally scarce but more numerous elsewhere, . . ., or rarely encountered due to secretive behavior or inaccessible habitat" (p. 76). From a spatial perspective, birders may focus on regions with multiple endemics species, given that they cannot be seen elsewhere and that they are typically rare. For rare species with a larger range, birders wanting to see these species may require a spotting network to locate them. Birders wanting to increase their life list for a particular region may concentrate their birding activities on the edges of those regions hoping to list species that are normally found in adjacent areas, but may fly into their focus region. From a temporal perspective, birders may feel a sense of urgency to see species before they become more endangered (requiring more effort or luck to see) or extinct (precluding the possibility of viewing). Birders also have arbitrary time constraints for listing their viewing accomplishments (e.g. number of species in a year, limited leisure time). From an ecological perspective, birders may seek out species facing significant threats, knowing that populations of those species will likely decline in the future. Moreover, birders may focus on threatened habitats which have a high number of rare species in order to optimize travel time and money. More positively, birders may contribute resources to help alleviate threats of rare species that they want to see in the future.

## Birder motivations to see rare birds

Many rare bird tourism attractions illustrate the range of issues associated with birders' desire to view or maintain rare species. Ecologically, the presence of tourists at nesting sites of the rare Yellow-eyed Penguin (*Megadyptes antipodes*) results in lower fledgling weights than at sites with no tourists, which would negatively impact survival (McClung *et al.* 2004). If tourism to view Royal Albatross (*Diomedea epomophora*) at Taiaroa Head, New Zealand, continues to rise, an increase in appropriate management is necessary to avoid negative impacts on both visitor experiences and albatross populations (Higham 1998). Research on birding tourism at Bald Eagle (*Haliaeetus leucocephalus*) viewing sites has led to improved management strategies that minimize negative ecological impact on the eagles (Anthony *et al.* 1995). Economically, Munn (1992) demonstrated that birding tourism to see various Macaw species (*Ara spp.*) in Peruvian jungle lodges generates between US$750 and US$4,700 in tourism revenues. Stoll and Johnson (1984) estimated the option price and existence value of Whooping Cranes (*Grus americana*) in the USA to be US$1.58 billion.

Birders have many different motivations to participate in their activities. Motivations involve a complex set of internal and external processes that direct one's behavior to achieve a certain goal. The opportunity to see rare birds is clearly an attraction for some birders, and rarity in birds can be influential for many other birder motivations in a tourism context. This section will review the research on birder motivations with an exploration of areas of overlap with rarity in birds.

Scott *et al.* (1999) found that key motivations for participating in a bird watching competition relate to achievement (e.g. being the first to spot a rare bird species), affiliation (e.g. strengthening of interpersonal relationships and enjoyment of being with others with similar interests), and appreciation (e.g. enjoyment of natural environment, relaxation, and solitude). While competition and the desire to see rare birds ranked lower than other motivational categories, the desire to participate influences decisions and results surrounding achievement and affiliation motivations. For example, participants in a birding competition had longer life lists than serious or advanced birders in other studies (Scott *et al.* 1999). Similarly, McFarlane's (1994) study of birder motivations in Alberta found that achievement factors (25 percent) rated lower than conservation (42 percent) or nature appreciation (33 percent). Nevertheless, the "search" is an important part of bird watching activities (Scott *et al.* 1999), a process which is intensified for rare species.

Sali and Kuehn (2006) classified the motivations of non-residential birders as emotional (e.g. enjoying outdoors, natural environment, birds), intellectual (e.g. studying, identification skills, sharing knowledge, teaching), physical (e.g. getting physical exercise, taking bird photos, traveling), social (e.g. competition with others, interacting with people with same interest), and spiritual (e.g. connecting with nature, renewing spiritual self). Emotional motivations were noted most highly by respondents. Rare birds can influence several of these motivations (e.g. emotional investment planning a search for rare species, intellectual challenge of identifying rare species, physical demands of locating rare species, social comparisons of life lists which include rare species). Kellert (1985) outlines five categories of motivations for casual (less specialized) and committed (more specialized) bird watchers: aesthetic (58 percent for casual, 28 percent for committed birders), fascination with birdlife (25: 53), closeness to nature (10: 11), identify as many species as possible (0: 3), and to be with family/friends (7: 5). Competition to increase life lists, which implies searching for rare species, is clearly the focus for a small portion of committed birders.

Unlike most other tourist types for whom escape is a critical motivation, seeking personal rewards (7.4 out of 9) was the dominant motivation for birders who were part of a nature-based tourism vacation experience (Partridge and MacKay 1998). In contrast, birders rated escaping personal environments (4.8) much lower. Some of these personal rewards include the chance of seeing rare species. For example, in the southeastern United States, birders rated "species not previously seen", "many different species", and "rare or endangered species" higher than hunters, in terms of relative importance on a wildlife viewing experience (Leuschner *et al.* 1989). Birding tourists in Australia wanted to see bird species which they have never seen (79 percent), are unique to Australia (67 percent), are vulnerable (63 percent), are rare (58 percent), and species which have not been seen by friends (23 percent; Green and Jones 2010). There was a strong correlation ($r^2 = 0.67$) between birders" willingness to join a birding tour and wanting to find difficult birds (Green and Jones 2010). Schneider and Salk (2002) show the importance of rare species to Minnesotan birders and wildlife viewers. From a list of 12 choices, warblers (*Dendroica spp.*, many of which are rare) rated 4.3 on a scale of 1–5 (with the 5 being the highest) and bald eagles (while increasing in numbers, are still perceived to be rare) rated 4.2.

Rare birds play a role in developing other kinds of motivations, such as achieving social status (e.g. having the longest life list) within a birding community (Liep 2001). Seeing rare species can help birders accumulate social capital, which gives one a bragging right or legendary status among peers (Cocker 2001; Connell

2009). Finding a rarity is "proof of ability", and repeating that process elevates one's status even further (Cocker 2001: 149). The extent to which social status might be enhanced by being the last to see a species before it becomes extinct is unknown, given that the process of declaring a species extinct can take years or decades. Moreover, developing a long life list, for which seeing rare species helps differentiate highly committed birders from less committed birders, can also elevate a birder's status. For example, birders with long life lists are described as "legends" (Liep 2001). In fact, several regions or countries have a club (e.g. Sweden: Club 300 Foundation for Bird Protection 2003) to which participants can join if they have seen, or desire to see, a substantial proportion of the species (usually involving a round number) that are possible in that area (Liep 2001). A committed top birder regularly accesses bird sighting information and is ready to travel at short notice to see rare birds, all in order to add to his/her life list and to maintain the title.

There are many other factors affecting birder motivations, especially as they relate to rare birds. First, related to gender, females may focus less on rare birds since they are reportedly less competitive, take fewer birding trips, and are less likely to use advanced optic or recording equipment than men (Sali and Zhang 2008; Scott *et al.* 2008). Second, related to specialization, advanced birders were more motivated by achievement motivations than were less specialized birder groups (McFarlane 1994). Appreciative and affiliation motivations decreased with increasing specialization, while conservation motivations were highest among novice birders (McFarlane 1994). Turpie and Ryan (1998) found that more specialized birders are more likely to keep lists and are willing travel further than less specialized birders. Thus, more specialized birders have different needs for educational programs and facilities than less specialized birders (Maple *et al.* 2010). Third, motivations vary significantly among sub-populations from different birding locations (Eubanks *et al.* 2004); every location has its own unique clientele. Fourth, distance seem to attract birders to birding locations (Mark 1981), unlike most other types of tourist attractions, for which distance is a deterrent. The opportunity to see novel or rare bird species may play a role in overcoming the significant constraints of distance.

## Bird rarity and the birding tourism industry

Rarity plays a significant role in marketing of the birding tourism industry. Advertising, rare bird alerts, and bird finding guides often refer to rare birds in order to influence birder behaviors. This section will review a few examples of each type. First, in describing their itineraries, birding tour companies regularly refer to rare or endangered species in order to attract birding customers. For example, for the Field Guides Birding Tours (2011) trip to southern France (italics added):

> We'll combine first-rate birding with a series of walks amid stunning mountain scenery, watching for the *endangered* Lammergeier or Bearded-Vulture (*Gypaetus barbatus*), the stunning but *elusive* Wallcreeper (*Tichodroma*

> *muraria*), Short-toed Eagle (*Circaetus gallicus*), European Honey-Buzzard (*Pernis apivorus*), Yellow-billed Chough (*Pyrrhocorax graculus*), Alpine Accentor (*Prunella collaris)*, Rock Bunting (*Emberiza cia*), and Citril Finch (*Carduelis citrinella*), among others. Come and see what awaits us this year!

One of the most sought after species in North America is the rare Kirtland's Warbler (*Dendroica kirtlandii*), especially on its breeding grounds in Michigan, a site visited by many birding tour companies. According to Fagan (2010) of the Field Guides Birding Tour company:

> It is listed as a federally endangered species and Near Threatened by BirdLife International. Nearly its entire population (current estimates of more than 3600 individuals) breeds in north-central Michigan in young Jack Pine forests. These forests were once naturally fire-maintained ecosystems, but they are now heavily managed through controlled burns and harvesting.

Similarly, the Victor Emmanuel Nature Tours (2011) birding tour to Churchill, Manitoba offers opportunities to see common species at "the most easily accessible tundra on this continent". Later, it focuses on the chances of seeing a specific rare species; such a possibility is sufficient to attract birders to these tours:

> For several decades it was famous as simply *the* place to find Ross's Gull (*Rhodostethia rosea*). However, this highly sought gull has been elusive in recent years, because the number of individuals declined to just a few. We will do our best to locate one, but chances are now slim.

Second, rare bird alerts are commonly used by birders to efficiently obtain information about rare birds in an area. Most often, local birders in the area pass on information about recent sightings that are noteworthy for other birders. These alerts are recorded on a phone message, displayed on a website, sent as an email message, indicated with a pager, or sent via other social media. The submissions typically report the date, observer, species seen, abundance status, location, and any logistical details for finding the site. Many of the species listed on the rare bird alerts are indeed rare, but also include uncommon migrants, species beyond their normal ranges, species in its range but at an unexpected time, and popular species. The rare bird alerts and informal information networks are extremely helpful to birders wanting to see rare birds. Still *et al.* (2004) suggest that, due to the nomadic nature of the birding community, birders in particular utilize networked communication enabled by web-based and mobile technologies.

With that information, serious birders commit considerable amounts of time and money to chasing rare birds. Some rare bird alerts list birding achievements, such as when a new species helped them reach a milestone in their life lists. A few examples indicate the importance of the North American Rare Bird Alert (2011):

- On May 14, 2010, Keenan Ennis reached his ABA #675 thanks to the Bar-tailed Godwit (*Limosa lapponica*) at Brigantine (Forsythe NWR), NJ.
- Larry Peavler saw a Yellow-breasted Bunting (*Emberiza aureola*) in a bone yard in Gambell, AK on September 2, 2009. It was his ABA #850.
- Jim Haw saw the Green-breasted Mango (*Anthracothorax prevostii*) in Beloit, Wisconsin, in 2007 for his 750th North American lifer – for the second time. He had reached 750 with a Little Stint (*Calidris minuta*) in Kentucky in August 2006, but then dropped back to 749 when the Yellow-chevroned Parakeet (*Brotogeris chiriri*) was delisted.

Third, bird finding guide books also influence birder behaviors by giving them information about what birds to expect in a given area, their relative abundance, and where to find them. In the *Costa Rica: Ecotravellers' Wildlife Guide*, Beletsky (1998: 2) describes how

> animal species and wild habitats are disappearing or deteriorating at an alarming rate. Because of the increasing emphasis on the importance of the natural environment by schools at all levels and the media's continuing exposure of environmental issues, people have enhanced appreciation of the natural world and increased awareness of environmental problems globally. They also have the very human desire to want to see undisturbed habitats and wild animals before they are gone, and those with the time and resources increasingly are doing so.

A less direct example comes from the bird-finding guide to Canada. It describes what birds to expect, for instance, in the southern Okanagan Valley of British Columbia. Most notable are several species which are of least concern on the IUCN Red List (IUCN 2010), but are rare or of special concern provincially, such as the Sage Thrasher (*Oreoscoptes montanus*) and Canyon Wren (*Catherpes mexicanus*) (Edwards and Davidson 2000). Similarly, Lentz (1989: 13) describes the rare and unique birds of Arizona:

> Birding is fantastic. Unique to Arizona alone are at least 25 species of North American birds, and several more are resident only in southern New Mexico and west Texas as well. The specialty birds you will want to see are numerous. From owls to nightjars and from hummingbirds to warblers, tanagers and orioles, a southeastern Arizona checklist reads like a birder's dream.

## Theorising birding motives for rare species

The human response to rarity requires some explanation in the form of theories and models that offer insight into the complex interactions among birder motivations, bird populations, society, and leisure. This section provides an overview of just five theories.

First, Iso-Ahola's (1989) *motivation theory* uses social psychology to explain two dimensions of leisure and tourism motivations. The first involves escaping personal (e.g. need for rest or escape from stressful situations) or interpersonal (e.g. away from family or work) environments, while the second involves seeking personal rewards (e.g. goals at the destination) or interpersonal environments (e.g. meet people with similar interests). Most general tourists are motivated more by the escape dimensions rather than the seeking dimensions (Iso-Ahola 1982), but Partridge and MacKay (1998) found the opposite for birders. In their study, birders were primarily motivated by seeking knowledge (e.g. "I look for educational experiences; I want to test my abilities; I want to develop my skills") and least motivated to escape personal environments (e.g. "I want rest and relaxation; I want to escape my personal problems"). The emphasis on personal rewards may relate to birders' specific goals, in the context of adding rare species to their life list (Partridge and MacKay 1998).

Second, Veblen's (2007) *theory of leisure class* suggests that people struggle "for status to elevate their social position in the eyes of others" (Scott 2010: 288). In trying to achieve higher status, people emulate those with higher status, in terms of their attitudes and activities. Historically, the display of status was achieved through two ways: conspicuous leisure and conspicuous consumption (Scott 2010). According to Scott (2010: 292), Veblen would understand that:

> Serious birders are motivated by standards of emulation that lead to patterns of conspicuous consumption. Seeing birds others have not seen is a badge of honor. Simultaneously, a life list remains a point of comparison among birders and compels them to travel long distances and spend large sums of money to list new birds. Birders are not different from other recreationists. Standards of emulation exist across a whole range of pastimes and inspire participants to spend lavishly as they seek conquests and enhanced reputations.

Some methods for birders to express this consumption, boost self-esteem, enhance ego, and promote self-recognition are to publish their accomplishments (e.g. life lists, year lists, and state/province lists by the American Birding Association 2010 and Surfbirds 2011), write about their listing adventures (e.g. Vardaman 1980;, Obmascik 2004; Dooley 2005), participate in bird listing competitions (e.g. World Series of Birding 2011), or talk with others about their travel experiences (Partridge and MacKay 1998).

Third, Sheard (1999) analyzes birding and twitching using Elias's (1978, 1982) *civilizing process theory*. Civilized processes refer to the improvement of social standards and behaviors (e.g. more refined manners, self-control, reduced violent or undesirable behaviors). Original bird watching activities required killing birds to identify, study, and draw birds. With the development of new technology (bird identification books, binoculars, scopes, and sophisticated cameras), Elias suggests that civilizing processes started, as watching and photographing birds became a substitute for killing birds. Moreover Sheard (1999: 182) suggests that:

> birdwatching developed as a more "civilized" substitute for hunting, but that a more "energetic," "sport-like" type of bird watching, birding, developed out

> of the originally more sedate bird watching, and that an even more sport-like activity, twitching, emerged as a response to the routinization and controlling of birding that has characterized the last 40 years.

The competitive elements of birding and twitching increase the desire see rare species in order to enlarge one's life list. Birding is considered to be a mimetic activity (Elias and Dunning 1986a, 1986b) or a leisure activity which appeals "directly to people's feelings and arouses them in various ways, eliciting excitement resembling that produced in real life situations but without its risks and dangers" (Sheard 1999: 184). Moreover, "bird watching is a mimetic activity par excellence that not only developed out of real sport, hunting, but continues to replicate and reproduce some of the pleasures and skills of hunting in symbolic and more civilized ways" (Sheard 1999: 185).

Fourth, *bioeconomic theory* examines perceived rarity of a wildlife population in the context of differing responses of demand by consumers and supply by wildlife providers (Hall *et al.* 2008). Applied to birding tourism, birders consume birding opportunities while host communities, protected areas, and tour guides supply the birding opportunities. Bioeconomic theory suggests that people value rare items; birders especially value the experience of sighting rare bird species. When a conservation organization lists an endangered species, that official notification about rarity generates more interest in that species. As a result, consumer demand can increase "if consumers believe that this is their last chance to obtain specimens before the species goes extinct" (Hall *et al.* 2008: 78). In some cases, increased exploitation (e.g. constant viewing of a rare species or development around its habitat) can cause the population size to decrease; with a smaller population, demand will increase, causing the price to increase. Normally, a higher price would reduce demand (Courchamp *et al.* 2006) until an equilibrium was reached.

Last, and in contrast to the previous theory, the *anthropogenic allee effect* may explain how a higher price can contribute to continued exploitation (Courchamp *et al.* 2006). This theory assumes a positive correlation between species rarity and its value to humans, which can then generate enough demand for the species that the market price is greater than the costs of finding and seeing the species. Thus, in certain cases, more people want those experiences of viewing rare species because they suggest higher social status (in the form of money, power, skill, and endurance), thus generating more demand and a resulting decline in the population of the target species (Hall *et al.* 2008). At the extreme, increased consumption of rare species results in a smaller population size, which drives demand up more. At some point the increased demand and resulting exploitation may drive the species to extinction. This situation has "serious implications for the conservation of species that are rare or that may become so, be they charismatic and emblematic or simply likely to become fashionable for certain activities" (Courchamp *et al.* 2006: 2405). Hall *et al.* (2006) describe several ecotourism activities in which wildlife viewing can generate such negative impacts on the target species (e.g. whale watching in British Columbia, Canada and Capercaillie [*Tetrao urogallus*] viewing in Scotland).

## Managing rare bird tourism

The emphasis of birders on rare birds raises many issues, both positive and negative, regarding bird conservation, science, and sustainability (e.g. see Buckley 2009; Hvenegaard 2009). On the one hand, rare bird tourism has the potential to positively impact birds and their habitats. A few examples will illustrate this. First, birders traveling to remote locations to view rare birds can spend considerable amounts of money in local communities (Hvenegaard *et al.* 1989). Those local communities and agencies responsible for wildlife may, in turn, work to conserve those birds and their habitats to maintain or increase economic benefits. Unfortunately, considerable leakage occurs when revenue leaves the local community (Sandbrook 2010). Second, birders who reliably record rare species in remote locations and share sightings with relevant agencies contribute valuable information to science and to efforts to help conserve those species. Third, the volume and continuity of birders near those rare birds may provide a certain level of security around them in the context of habitat protection or reduced exploitation. Fourth, many birders volunteer their skills and time for research and monitoring (McFarlane and Boxall 1996). For example, through "volunteer tourism", birders help scientists conduct research on rare birds (e.g. Earthwatch). Birders also conduct annual breeding bird surveys that provide valuable information for bird conservation efforts (McFarlane and Boxall 1996). Fifth, many birders donate money to help manage or conserve critical bird habitat for endangered species (McFarlane and Boxall 1996; Hvenegaard 2002).

On the other hand, rare bird tourism can negatively impact birds and their habitats (Buckley 2004). Higham and Lück (2007) describe many paradoxes associated with ecotourism, most of which are unresolved. Some of these negative impacts are described here. First, birders who discover a rare species often tell others about their sightings (e.g. word of mouth, rare bird alerts, published articles) which leads to more people wanting to see the species. Enough people viewing the bird onsite can lead to disturbance and habitat change. Second, many birders use carbon-based modes of transportation in their extensive travels to see rare species (Simmons and Becken 2004). The increase in greenhouse gases contributes to climate change which has a negative impact on species and their habitats. Like polar bear viewers at Churchill, Manitoba (Dawson *et al.* 2010), some birders respond by purchasing carbon offsets for their travel or by reducing travel substantially. Third, marketing efforts may use "greenwashing" to describe bird watching tourism as beneficial or neutral, when it is actually detrimental to the natural environment, or bird habitats. Fourth, marketing can create a sense of urgency to see a rare species before it becomes extinct simply to generate additional revenue.

Many management and planning efforts have sought to address these issues to improve the positive impacts and decrease the negative impacts of birding tourism. Research has helped identify, manage, and resolve some issues. While a full examination is explored in other publications (Norris and Pain 2002; Manfredo 2002; Buckley 2004; McCool 2009), a few points deserve mention. From a conservation perspective, securing native habitat is critical for rare birds (e.g. protected areas and habitat management). As Shelton's case study on the Kakapo (*Strigops habroptilus*) in New Zealand reveals (see Lemieux and Eagles this volume), in some exceptional cases, species with very low populations can be translocated to new habitats or bred in captivity with a goal of eventual release back into the wild (Bell and Merton 2002). Captive birds, however, are not acceptable for birding lists (American Birding Association 2006). From an educational perspective, some organizations encourage birders to look at rare birds for more than adding a species to a life list; birders should also recognize other important aspects of these rare bird species (e.g. behavior, song, habitat, conservation status). In addition, many birding and nature organizations encourage birders to abide by a code of ethics or birding etiquette (American Birding Association 2006), which stresses minimal impacts on rare species. An ethical approach is needed for all ecotourism activities (Nowaczek *et al.* 2007). From a technical perspective, it is now possible to remotely view rare species (e.g. web cam located near a nesting site; Chambers 2007), but according to listing protocols, such sightings may not count toward a birding list. From an experiential perspective, research can help clarify dimensions of quality wildlife viewing (Anderson *et al.* 2010), the dynamics of birding tours (Che 2003), the role of birding tour leaders (Curtin 2010), and the value of certification (Honey 2008).

## Conclusion

Birding for rare bird species illustrates several important issues related to last chance tourism (Lemelin *et al.* 2010). First, since many birders desire to see as many species as possible, often through a listing process, birders place a disproportionate level of emphasis on rare species. Second, since rare species have small or declining populations or significant threats to their long-term survival, birders value these species and the opportunity to experience them. Third, once recorded by a birder, a rare bird species can never be removed from a birder's "list", even if that species becomes extinct. Last, birding activities focused on rare birds can cause subsequent negative impacts on those populations and habitats (Dawson *et al.* 2009).

Thus, the possibility of seeing rare bird species is tremendously attractive to many birders. Regardless of a birder's motive or use of a bird (e.g. viewing, study, photography, or listing), the interaction between the birder and the bird raises important questions. Should the interaction even be allowed, given the rarity of the species, uncertainty about impacts, and the risk of extinction? Should information about the potential for viewing the rare bird be publicized? How should the interaction be managed? How can the interaction be used to promote the long-term conservation of the species? Can the behavior of birders be managed to minimize negative impacts? Can and should birding organizations generate a birding culture that does not equate self-worth with the size of one's life list?

Potential answers to these questions require a careful analysis of past practices, problem-based research, individual and collective birding behavior, site management capacity, and more. Integrative solutions will also require the involvement of all stakeholders in the birding tourism industry, including birding tour companies, guides, protected area managers, researchers, birding organizations, endangered species scientists, local communities, environmental nongovernmental organizations, relevant government agencies, and the public. Given the rarity of the target species, and the resulting risk of an irreversible impact, any solution should employ a precautionary approach to decision making. The over-riding questions to ask should be "How can birding for rare species minimize impacts on that species and its habitats?" and "How can rare bird tourism be managed to promote conservation benefits to the target species?"

## Acknowledgments

I thank the Social Sciences and Humanities Research Council of Canada for funding, the University of Alberta's Augustana Campus for logistical support, Nhial Tiitmamer Kur for research assistance, and the editors and an anonymous reviewer for feedback.

## References

Addison, A.C. (2008) *Disappearing World: 101 of the Earth's Most Extraordinary and Endangered Places*, New York: Harper Collins.

American Birding Association (2003) "American Birding Association principles of birding ethics", *2003 ABA Member Handbook*, Colorado Springs, CO: American Birding Association, p. 24.

American Birding Association (2006) "Listing basics". Online: www.aba.org/bigday/listing.html (accessed 6 March 2011).

American Birding Association (2010) "ABA big day and list report". Online: www.aba.org/bigday/ (accessed 6 March 2011).

Anderson, L.E., Manning, R.E., Valliere, W.A. and Hallo, J.C. (2010) "Normative standards for wildlife viewing in parks and protected areas", *Human Dimensions of Wildlife* 15: 1–15.

Anthony, R.G., Steidl, R.J. and McGarigal, K. (1995) "Recreation and bald eagles in the Pacific Northwest", in R.L. Knight and K.J. Gutzwiller (eds.) *Wildlife and Recreationists: Coexistence Through Management and Research*, Washington, DC: Island Press, pp. 223–242.

Becken, S. and Hay, J. (2007) *Tourism and Climate Change: Risks and Opportunities*, Clevedon, UK, and Buffalo, NY: Channel View Publications.

Beletsky, L. (1998) *Costa Rica: The Ecotravellers' Wildlife Guide,* San Diego, CA: Academic Press.

Bell, B.D. and Merton, D.V. (2002) "Critically endangered bird populations and their management", in K. Norris and D.J. Pain (eds.) *Conserving Bird Biodiversity: General Principles and their Application,* Cambridge, MA: Cambridge University Press.

BirdLife International (2008) *State of the World's Birds: Indicators for our Changing World*, Cambridge: BirdLife International.

Brennan, L.A. and Kuvlesky, W.P. (2005) "North American grassland birds: an unfolding conservation crisis?", *Journal of Wildlife Management* 69(1): 1–12.

Bryan, H. (1977) "Leisure value systems and recreational specialization: the case of trout fishermen", *Journal of Leisure Research* 9(3): 174–187.

Buckley, R. (2004) "Impacts of ecotourism on birds", in R. Buckley (ed.) *Environmental Impacts of Ecotourism*, Cambridge, MA: CABI.

Buckley, R. (2009) "Evaluating the net effects of ecotourism on the environment: a framework, first assessment and future research", *Journal of Sustainable Tourism* 17(6): 643–672.

Carver, E. (2009) *Birding in the United States: A Demographic and Economic Analysis*, Arlington, VA: US Fish and Wildlife Service.

Chambers, C.N.L. (2007) "'Well it's remote, I suppose, innit?' the relational politics of bird-watching through the CCTV lens", *Scottish Geographical Journal* 123(2): 122–134.

Che, D. (2003) "Guided birding tours: an examination of the market, important tour parameters, and participant demographics", *Proceedings of the 2003 Northeastern Recreation Research Symposium*, Gen. Tech. Rep. NE-317, Newton Square, PA: US Department of Agriculture, Forest Service, Northeastern Research Station, pp. 194–202.

Club 300 Foundation for Bird Protection (2003) "The Club 300 Foundation for Bird Protection". Online: www.club300.se/Birdprot/Birdprotection.aspx (accessed 6 March 2011).

Cocker, M. (2001) *Birders: Tales of a Tribe*, London: Jonathan Cape.

Connell, J. (2009) "Bird watching, twitching and tourism: towards an Australian perspective", *Australian Geographer* 40(2): 203–217.

Conservation International (2011) "The biodiversity hotspots". Online: www.conservation.org/explore/priority_areas/hotspots/Pages/hotspots_main.aspx (accessed 6 March 2011).

Courchamp, F., Angulo, E., Rivalan, P., Hall, R.J., Signoret, L., Bull, L. and Meinard, Y. (2006) "Rarity value and species extinction: the anthropogenic allee effect", *PLoS Biology* 3(12): 2405–2410.

Curtin, S. (2010) "Managing the wildlife tourism experience: the importance of tour leaders", *International Journal of Tourism Research* 12: 219–236.

Dawson, J., Lemelin, H. and Stewart, E. (2009) "Last chance tourism: risks and opportunities of an emerging market", Conference Proceedings for the Travel and Tourism Research Association's Canada 2009 Conference, Guelph, Ontario.

Dawson, J., Stewart, E.J., Lemelin, R.H. and Scott, D. (2010) "The carbon cost of polar bear viewing tourism in Churchill, Canada", *Journal of Sustainable Tourism* 18(3): 319–336.

Dooley, S. (2005) *The Big Twitch: One Man, One Continent, A Race Against Time – A True Story About Birdwatching*, Crows Nest, NSW, Australia: Allen & Unwin.

Edwards, Y. and Davidson, G. (2000) "British Columbia", in J.C. Finlay (ed.) *Bird-Finding Guide to Canada,* Toronto: The Canadian Publishers.

Elias, N. (1978) *The Civilizing Process, Volume I: The History of Manners*, Oxford: Blackwell.

Elia, N. (1982) *The Civilizing Process, Volume II: State Formation and Civilization,* Oxford: Blackwell.

Elias, N. and Dunning, E. (1986a) "Leisure in the spare-time spectrum", in N. Elias and E. Dunning (eds.), *Quest for Excitement: Sport and Leisure in the Civilizing Process*, Oxford: Blackwell, pp. 91–125.

Elias N. (1986b) "The quest for excitement in leisure", in N. Elias and E. Dunning (eds.), *Quest for Excitement: Sport and Leisure in the Civilizing Process*, Oxford: Blackwell, pp. 63–90.

Eubanks, T.L., Stoll, J.R. and Ditton, R.B. (2004) "Understanding the diversity of eight birder sub-populations: socio-demographic characteristics, motivations, expenditures and net benefits", *Journal of Ecotourism* 3(3): 151–172.

Fagan, J. (2010) "The Kirtland's warbler". Online: fieldguides.com/news/?p=4174 (accessed 6 March 2011).

Field Guides Birding Tours. (2011) "France: Camargue & Pyrenees". Online: fieldguides.com/bird-tours/france (accessed 6 March 2011).

Green, R.J. and Jones, D. (2010) *Practices, Needs and Attitudes of Bird-watching Tourists in Australia*, Queensland, Australia: CRC for Sustainable Tourism.

Hall, R.J., Milner-Gulland, E.J. and Courchamp, F. (2008) "Endangering the endangered: the effects of perceived rarity on species exploitation", *Conservation Letters* 1: 75–81.

Higham, J. (1998) "Tourists and albatrosses: the dynamics of tourism at the Northern Royal Albatross Colony, Taiaroa Head, New Zealand", *Tourism Management* 19: 521–31.

Higham J. and Lück, M. (2007) "Ecotourism: pondering the paradoxes", in J. Higham (ed.) *Critical Issues in Ecotourism: Understanding a Complex Tourism Phenomenon*, London: Elsevier, pp. 117–135.

Honey, M. (2008) "Setting standards: certification programmes for ecotourism and sustainable tourism", in A. Stronza and W.H. Durham (eds.) *Ecotourism and Conservation in the Americas*, Cambridge, MA: CABI, pp. 234–269.

Hughes, H. (2008) *Frommer's 500 places to visit before they disappear*. Hoboken, NJ: Frommers.

Hvenegaard, G.T. (2002) "Birder specialization differences in conservation involvement, demographics, and motivations", *Human Dimensions in Wildlife* 7(1): 21–36.

Hvenegaard, G.T. (2009) "Potential conservation benefits of wildlife festival tourism", *Event Management* 13: S8–9.

Hvenegaard, G.T., Butler, J.R. and Krystofiak, D.K. (1989) "The economic values of bird watching at Point Pelee National Park, Canada", *Wildlife Society Bulletin* 17(4): 526–531.

IUCN (2010) "The IUCN Red List of Threatened Species". Online: www.iucnredlist.org/apps/redlist/search (accessed 6 March 2011).

Iso-Ahola, S.E. (1982) "Toward a social psychological theory of tourism motivation: a rejoinder", *Annals of Tourism Research* 9: 256–262.

Iso-Ahola, S.E. (1989) "Motivation for leisure", in E.L. Jackson and T.L. Burton (eds.), *Understanding Leisure and Recreation: Mapping the Past, Charting the Future*, State College, PA: Venture Publishing Inc, pp. 247–279.

Kellert, S. (1985) "Bird watching in American society", *Leisure Sciences* 7(3): 343–360.

Lebbin, D.J., Parr, M.J. and Fenwick, G.H. (2010) *The American Bird Conservancy Guide to Bird Conservation*, Chicago: University of Chicago Press.

Lemelin, H., Dawson, J., Stewart, E.J., Maher, P. and Lueck, M. (2010) "Last-chance tourism: the boom, doom, and gloom of visiting vanishing destinations", *Current Issues in Tourism* 13(5): 477–493.

Lentz, J.E. (1989) *Great Birding Trips of the West*, Santa Barbara, CA: Capra Press.

Leuschner, W.A., Ritchie, V.P. and Dean, F.S. (1989) "Opinions on wildlife: responses of resource managers and wildlife users in the Southeastern United States", *Wildlife Society Bulletin* 17(1): 24–29.

Liep, J. (2001) "Airborne Kula: the appropriation of birds by Danish ornithologists", *Anthropology Today* 17(5): 10–15.

MacDonald, G.M. (2003) *Biogeography: Space, Time, and Life*, New York: John Wiley & Sons.

Manfredo, M.J. (ed.) (2002) *Wildlife Viewing: A Management Handbook*, Corvallis, OR: Oregon State University Press.

Maple, L.C., Eagles, P.F.J. and Rolfe, H. (2010) "Birdwatchers" specialization characteristics and national park tourism planning", *Journal of Ecotourism* 9: 219–238.

Mark, D. (1981) "On the positive relation between distance and attractivity in recreation travel: the example of birding", *Ontario Geography* (17): 43–90.

McClung, M.R., Seddon, P.J., Massaro, M. and Setiawan, A.N. (2004) "Nature-based tourism impacts on yellow-eyed penguins *Megadyptes antipodes*: does unregulated visitor access affect fleding weight and juvenile survival?", *Biological Conservation* 119: 279–85.

McCool, S.F. (2009) "Challenges and opportunities at the interface of wildlife-viewing marketing and management in the twenty-first century", in M.J. Manfredo, J.J. Vaske, P.J. Brown, D.J. Decker and E.A. Duke (eds.) *Wildlife and Society: The Science of Human Dimensions*, Washington, DC: Island Press.

McCracken, J.D. (2005) "Where the Bobolinks roam: the plight of North America's Grassland Birds", *Biodiversity* 6(3): 20–29.

McFarlane, B.L. (1994) "Specialization and motivations of birdwatchers", *Wildlife Society Bulletin* 22(3): 361–370.

McFarlane, B.L. and Boxall, P.C. (1996) "Participation in wildlife conservation by birdwatchers", *Human Dimensions of Wildlife* 1(3): 1–14.

Munn, C.A. (1992) "Macaw biology and ecotourism, or 'when a bird in the bush is worth two in the hand'", in S.R. Beisssinger and N.F.R. Snyder (eds.) *New World Parrots in Crisis: Solutions from Conservation Biology*, Washington, DC: Smithsonian Institution Press, pp. 47–72.

Newton, I. (2003) *The Speciation and Biogeography of Birds*, Amsterdam: Academic Press.

Norris, K. and Pain, D.J. (eds.) (2002) *Conserving Bird Biodiversity: General Principles and their Application*, Cambridge: Cambridge University Press.

North American Rare Bird Alert (2011) "Report a rare bird". Online: www.narba.org/default.aspx/MenuItemID/104/MenuGroup/Home.htm (accessed 6 March 2011).

Nowaczek, A.M., Moran-Cahusac, C. and Fennell, D.A. (2007) "Against the current: striving for ethical ecotourism", in J. Higham (ed.) *Critical Issues in Ecotourism: Understanding a Complex Tourism Phenomenon*, London: Elsevier, pp. 136–157.

Obmascik, M. (2004) *The Big Year: A Tale of Man, Nature, and Fowl Obsession*, New York: Free Press.

Oddie, B. (1995). *Bill Oddie's Little Black Bird Book*, 2nd edn, London: Robson.

Partridge, C. and MacKay, K.J. (1998) "An investigation of the travel motives of birdwatchers as a nature-based tourist segment", *Journal of Applied Recreation Research* 23(3): 263–287.

Pimm, S., Raven, P., Peterson, A., Sekercioglu, C.S. and Ehrlich, P.R. (2006) "Human impacts on the rates of recent, present, and future bird extinctions", *Proceedings of the National Academy of Sciences* 103(29): 10941–1046.

Price, J. (2003) "A birder's guide to climate change", *Birding* 35: 630–39.

Sali, M.J.G. and Kuehn, D.M. (2006) "Gendered-based motivations of non-residential birdwatchers in New York State: qualitative study", in Burns, R. and Robinson, K. (eds.) *Proceedings of the 2006 Northeastern Recreation Research Symposium*, Newton Square, PA: Department of Agriculture, Forest Service, Northern Research Station.

Sali, M.J.G. and Zhang, L. (2008) "Motivations for male and female birdwatchers in New York State", *Human Dimensions of Wildlife* 13(3): 187–200.

Sandbrook, C.G. (2010) "Local economic impact of different forms of nature-based tourism", *Conservation Letters* 3: 21–28.

Schaffner, S. (2009) "Environmental sporting: birding at superfund sites, landfills, and sewage ponds", *Journal of Sport & Social Issues* 33(3): 206–229.

Schneider, I.E. and Salk, R. (2002) *Information Needs and Experience Preferences of Birders and Watchable Wildlife Participants*, St. Paul, MN: University of Minnesota, Department of Forest Resources.

Scott, D. (2010) "What would Veblen say?", *Leisure Sciences* 32: 288–294.

Scott, D., Baker, S.M. and Kim, C. (1999) "Motivations and commitments among participants in the Great Texas birding classic", *Human Dimensions of Wildlife* 4(1): 50–67.

Scott, D., Moore, R.L. and Lee, J-H. (2008) "Role of gender and level of specialization in predicting motives among birdwatchers", *Book of Abstracts of the Papers* presented at the Twelfth Canadian Congress on Leisure Research, Montréal, PQ: Applied Human Sciences Department, Concordia University, pp. 394–398.

Sekercioglu, C.H., Daily, G.C. and P.R. Ehrlich (2004) "Ecosystem consequences of bird declines", *PNAS* 101(52): 18042–18047.

Sheard, K. (1999) "A twitch in time saves nine: bird watching, sport, and civilizing processes", *Sociology of Sport Journal* 16: 181–205.

Simmons, D.G. and Becken, S. (2004) "The cost of getting there: impacts of travel to ecotourism destinations", in R. Buckley (ed.) *Environmental Impacts of Ecotourism*, Cambridge, MA: CABI.

Still, K., Vainamo, S. and Isomursu, M. (2004) "Connecting networked communities – case birdwatchers", *IADIS International Conference Web Based Communities*, pp. 27–34.

Stoll, J. and Johnson, L.A. (1984) "Concepts of value, nonmarket valuation and the case of the Whooping Crane", *Transactions of the North American Wildlife and Natural Resources Conference* 49: 382–392.

Surfbirds (2011) "Life and year list rankings". Online: www.surfbirds.com/rankings.html (accessed 6 March 2011).

Turpie, J.K. and Ryan, P.G. (1998) *The Nature and Value of Birding in South Africa*, Rondebosch: BirdLife South Africa.

Vardaman, J.M. (1980) *Call Collect, Ask For Birdman: The Record-Breaking Attempt to Sight 700 Species of North American Birds Within One Year*, New York: St. Martin's Press.

Veblen, T. (2007) *The Theory of the Leisure Class*, Oxford: Oxford University Press (original published in 1899).

Victor Emmanuel Nature Tours (2011) Churchill: Jun 12–18, 2011. Online: www.ventbird.com/birding-tour/2011/06/12/churchill (accessed 6 March 2011).

World Series of Birding (2011) "World series of birding". Online: www.birdcapemay.org/wsob.shtml (accessed 6 March 2011).

# 7 The transformation of polar bear viewing in the Hudson Bay region, Canada

*Emma Stewart, Jackie Dawson, and Raynald Harvey Lemelin*

## Introduction

Canada is home to approximately 15,500 of the estimated 20,000 to 25,000 polar bears in global polar regions (Environment Canada 2009). Given that 13 of 19 subpopulations of polar bears (*Ursus maritimus*) are found either wholly or partially in Canada, it is not surprising that a number of destinations along the coast of Hudson Bay, James Bay in Nunavik (Northern Quebec), Ontario, Manitoba, and Nunavut have developed tourism opportunities to view the three populations found in these areas (i.e., Western Hudson Bay, Southern Hudson Bay, Foxe Basin). As the self-declared 'polar bear capital of the world', Churchill in northern Manitoba is perhaps the most well-known place to view polar bears, but other viewing locations include the aptly named Polar Bear Provincial Park in Ontario and, to the north, Ukkusiksalik National Park in Nunavut. However, due to climate change it is widely thought that polar bears in this region (and elsewhere) are in a period of transformation with future prospects that are described as 'grim' (Vongraven 2009: 326). As Derocher *et al.* (2004: 173) indicates, polar bears are 'vulnerable' to habitat change induced by climate warming. The consequent *rush* of tourists to view polar bears in their natural environment, before they are gone for good, is perhaps one of the most obvious examples of the last chance tourism phenomenon (Lemelin, Dawson, Stewart, Maher, and Lück 2010a; Dawson *et al.* 2010).

Polar bears depend heavily on sea ice for both feeding and breeding activities therefore changes in sea ice distribution and abundance can have significant impacts on the health of the species (Stirling and Derocher 1993; Stirling and Parkinson 2006). For example, less time on sea ice will lead to less prey availability, increasing fasting periods and a significant decline in reproduction rates (Schliebe *et al.* 2008). Given the warming global climate has precipitated a significant decrease in sea ice extent and thickness in the Arctic regions (ACIA, 2004; Furgal and Prowse 2008), it is not surprising that by the mid-twenty-first century an estimated two-thirds of the world's polar bears will be locally extinct (Amstrup, Marcot, and Douglas 2007). Concerns for the well-being of this species has resulted in polar bears being listed on the *US Endangered Species Act* and listed as 'vulnerable' on the IUCN Red List of Threatened Species. The Western Hudson Bay subpopulation is listed as 'threatened' under the *Endangered Species*

*Act of Manitoba* and the Southern Hudson Bay subpopulation has been listed on the *Ontario Endangered Species Act*.

In the short term, an observed increase in demand for viewing polar bears, and an increased willingness to pay to see polar bears before they are gone is likely to positively influence revenues for tour operators, particularly for those working in established nature-based destinations such as Churchill (Dawson *et al.* 2010). However, in the long term, sustainability of the industry is of great concern, for without polar bears there is no polar bear viewing industry. Thus, the short-term boom from last chance tourism will, according to most long-term climate change predictions, end up unfavorably (Lemelin *et al.* 2010b). But for the region as a whole, it may not be all doom, especially if a localized shift in the use of onshore areas by the Western Hudson Bay polar bear population occurs. Other locations in Hudson Bay may be able to learn from the Churchill experience, develop their own polar bear viewing industry, and capitalize on what has become a multi-million dollar business.

Drawing on a number of empirical studies of nature-based tourism in the region, this chapter explores polar bear viewing in the Hudson Bay region, the emergence of Churchill as a last chance tourism destination, and the relationship between climate change, polar bears and projected shifts in habitat range. The discussion addresses the following two questions: what might happen to Churchill when all the 'last chances to see polar bears' have been exhausted, and, conversely, how might the wider Hudson Bay region respond? The way in which coastal communities and protected natural areas along Hudson Bay adapt to the future status of the polar bear can be considered a test-case of how last chance tourism destinations address and adapt to the multi-layered complexities of climate-induced change.

## Environmental context of the Hudson Bay region

Across the Hudson Bay region polar bear habitat spans the boreal forests of northern Ontario to the tundra of Fury and Hecla Strait between the Meville Peninsula and Baffin Island in Nunavut (Peacock, Derocher, Lunn, and Obbard 2010). There are three semi-discrete subpopulations of polar bears in the region including (1) Southern Hudson Bay (SHB) (2) Western Hudson Bay (WHB) and (3) Foxe Basin (FB) (see Figure 7.1). Population estimates indicate that SHB and WHB subpopulations are of similar size each around 900 individuals (Lemelin and Dyck 2008; Regehr, Lunn, Amstrup, and Stirling, 2007). Population size for polar bears in FB from 1989 to 1994 was estimated to be approximately 2,200 (2,197 ± 260 SE) (Sahanatien and Derocher 2006); more recent estimates are not available (Taylor, Lee, Laake, and McLoughlin 2006).

Hudson Bay undergoes a complete seasonal sea ice cycle each year, transitioning from fully ice covered in winter to open water in summer (Tivy, Alt, Howell, Wilson, and Yackel 2007). Currently, the region begins to freeze-up in November and is completely ice covered by January. Melt begins in early May and by August the whole region is ice-free until October (Falkingham, Chagnon, and McCourt 2001). During the ice-free period the range of the SHB polar bear

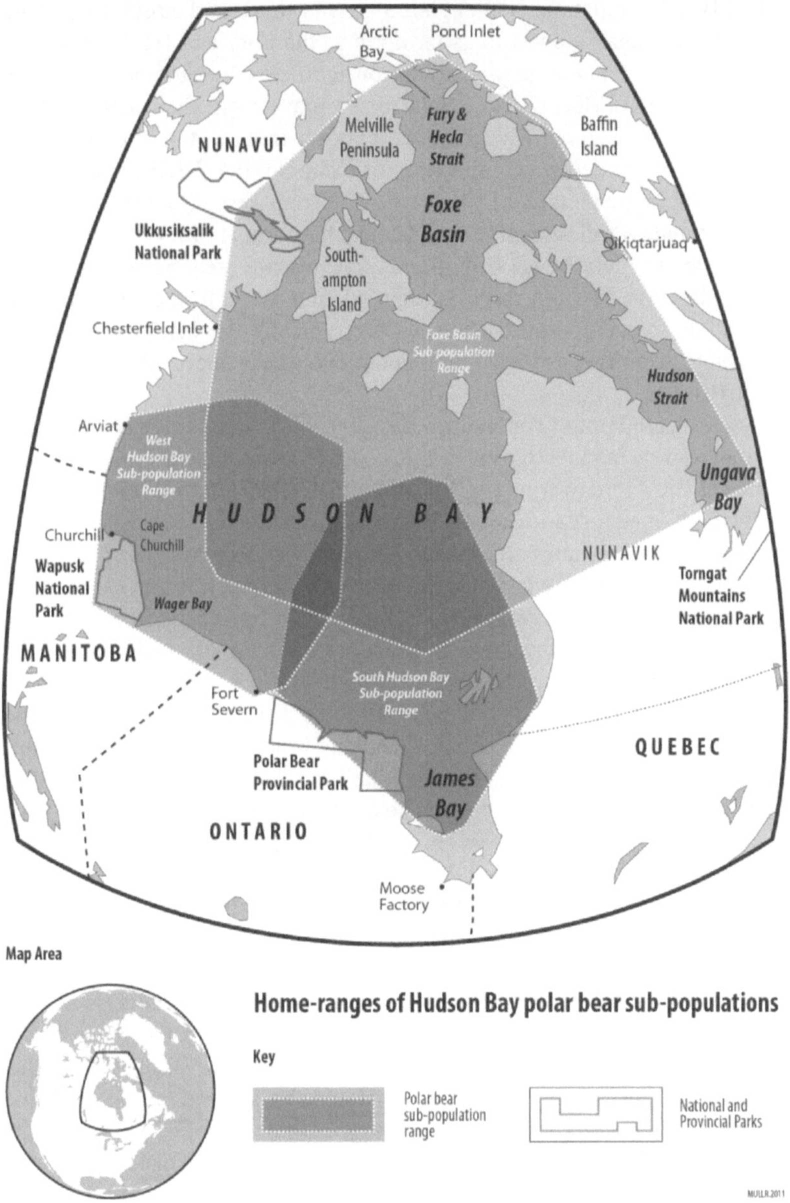


*Figure 7.1* Map of the Hudson Bay Region of Canada (Cartography by Shawn Mueller)

population is from the Manitoba–Ontario border east to Nunavik and south to James Bay. During this time bears can fast for up to five months. Bears belonging to the WHB subpopulation reside along the western shore of Hudson Bay north from the Manitoba–Ontario border to Chesterfield Inlet in Nunavut. The range of the FB subpopulation is thought to extend west and north from Hudson Strait and the Nunavik coast to as far north as the Fury and Hecla Strait. In this area polar bears can forage year round on ice floes. In anticipation of the sea ice freezing across Hudson Bay, polar bears move from their summer habitats towards coastal locations where the ice appears first, so that they can prey on seals and other ice dependent species. For example, the WHB population tends to congregate in the vicinity of Churchill, with some bears (largely subadults and some adult males) following the first sea ice that forms north of Churchill, while others move out onto the ice from the Cape Churchill area. Similarly, on Southampton Island, bears from the FB subpopulation move to the southern bays of the island as freeze-up approaches and are typically the last bears to return to the frozen Hudson Bay as freeze-up is about two weeks later in this area (Peacock *et al.* 2010).

The sea ice climatology of Hudson Bay reveals that the region has experienced an overall decline in sea ice cover over the past 35 years, a trend that is projected to continue as the Arctic warms (ACIA 2004; IPCC 2007). Parkinson and Cavalieri (2008) report annual sea ice extent within Hudson Bay is decreasing at –5.3 per cent per decade with summertime (July–September) decreases of up to –19.5 per cent per decade. Over the last 40 years there has been a trend to earlier summer break-up of the sea ice in western Hudson Bay. As a consequence, it is considered that sea ice breakup occurs, on average, about three weeks earlier than it did 30 years ago (Towns, Derocher, Stirling, and Lunn 2010). As sea ice is critical for the polar bears, the earlier breakup is believed to be a factor in the declining health of the regional polar bear population (Scott and Marshall 2010) with Stirling and Parkinson (2006) reporting that female polar bears in the Hudson Bay region may stop reproducing altogether within the next 20–30 years.

A 20 year data set (1984 to 2004) of the WHB subpopulation reveals a decline in population from approximately 1,200 to 935 respectively (Lunn, Stirling, Andriashek, and Kolenosky 1997; Regehr *et al.* 2007). According to Polar Bear Specialist Group (2011), the SHB subpopulation is currently considered to be more stable at 900 to 1000, although reductions in body condition have been detected which could signal potential decline in the SHB population (Obbard, McDonald, Howe, Regehr, and Richardson 2007). The last population survey indicated that there were approximate 1,677 to 2,717 polar bears in the Foxe Basin subpopulation. Reduced sea ice in the Foxe Basin has generated similar predictions of decline for this population as well (Stirling and Parkinson 2006). However, the ice-free season in the Foxe Basin is shorter than in the southern reaches of Hudson Bay, and ice is predicted to continue to form here in the winter, meaning this subpopulation is at less risk than those in the more southern regions of the Bay (Peacock *et al.* 2010). Towns *et al.* (2010) examined the capture locations of polar bears on land in western Hudson Bay over a 19-year period (1986–2004) to assess temporal trends in the distribution of the population and found that the distribution of bears had shifted northward and eastward.

## Polar bear viewing in the Hudson Bay region

The opportunities for visitors to view polar bears in the region has evolved as a result of Hudson Bay's unique sea ice climatology, which acts as a platform for the food sources that live there, the bear's related patterns of feeding, the population ecology of the species, as well the entrepreneurship displayed by local residents and operators (Peacock *et al.* 2010). Although, polar bear viewing opportunities exist throughout the Canadian Arctic from within protected natural areas as well as from communities such as Arctic Bay, Qikiqtarjuaq and Pond Inlet, Nunavut (Explore Nunavut 2011), the following section describes the main centers for viewing opportunities specifically in the Hudson Bay region.

### *Polar Bear Provincial Park (PBPP), Northern Ontario*

As one of Canada's largest protected areas, Polar Bear Provincial Park (PBPP) covering 23,552 km$^2$ is located in northern Ontario and was established in 1970 to protect important coastal, wetland and tundra ecosystems. As the name of the park suggests, polar bears are a key management focus of the area as it protects the world's most southerly population. Viewing of polar bears in Ontario is limited to a few local entrepreneurs (Wild Wind Tours, Polar Bear Expeditions, Tommy Miles from Fort Severn) from the three First Nation signatories (Attawapiskat, Weenusk First Nation at Peawanuck, Washaho Cree Nation at Fort Severn) (Lemelin *et al.* 2010c). The park's wilderness classification prohibits mechanized travel (ATVs and snowmobiles) for purposes other than traditional harvesting and some wildlife viewing by visitors. Therefore, all guided wildlife viewing activities occurring in the park are restricted to river corridors (i.e. the Sutton River) and along beach ridges where semi-permanent infrastructures exist (i.e. trappers cabins and prospector tents). Since the park is also defined as non-operational (no fees or staff on site, limited facilities, no formal management plan), current estimates of park visitors are unclear; however, indications from the operators would suggest only a few hundred people visit PBPP annually, but not all of them are polar bear viewing tourists (Lemelin and Dyck 2008).

### *Ukkusiksalik National Park (UNP), Nunavut*

Designated in 2003, Ukkusiksalik National Park covering 20,500 km$^2$ is located west of the community of Repulse Bay and surrounds Wager Bay, a 100 km long saltwater inlet on the northwest coast of Hudson Bay in Nunavut. Named after the soapstone found there, the park includes eskers, mudflats, cliffs, rolling tundra banks, and unique coastal regions (Explore Nunavut 2011; Parks Canada 2011). At Wager Bay's south shore there is a large polar bear denning area (pregnant polar bears use dens as birthing sites), and in July and at the beginning of August, polar bears from the FB subpopulation can be observed from boats on ice floes or on islands. The park is particularly renowned for opportunities to observe polar bears swimming. Operated by Frontiers North, Sila Lodge, a small eco-lodge on the shores of the Sila River is the main base camp for wildlife viewing where an estimated 130 polar bears are thought to live. According to Portman (2004), each year about 300 people travel to Sila Lodge, but discussions with Parks Canada officials could not corroborate these numbers since official visitation figures for UNP are not, as of yet, available. Common to other national parks in the Canadian Arctic, only Inuit and park wardens are permitted to carry guns in the protected area, so visits by unaccompanied tourists are discouraged, which severely curtails polar bear viewing activities (Maher and Lemelin 2011).

### *Churchill, Northern Manitoba*

Over four decades the community of Churchill in northern Manitoba has built a global reputation as being the *best* place to view polar bears in their natural habitat. When polar bears are forced ashore in summer by melting sea ice they largely fast, and then in the autumn generally congregate for approximately six weeks (starting in late October) along the western shore of Hudson Bay where they await the formation of sea ice (Stirling 1998). In essence, tourism operators have capitalized on this 'waiting period' to provide visitors with an opportunity to view polar bears in their natural habitat. The viewing of polar bears in this area is further facilitated by the habituation and tolerance of the species to human presence, and their curiosity and propensity to entertain wildlife viewers (Lemelin and Smale 2006).

In order to protect the tundra environments, two protected areas have been designated. Churchill Wildlife Management Area was created in 1978 to protect denning areas, goose nesting sites and caribou grounds, and Wapusk National Park was created in 1996 to represent the Hudson-James Lowlands natural region (Parks Canada 2007). The park protects important maternity denning areas south east of Churchill (Clark 2003). The majority of polar viewing activities take place along the narrow coastal strip east of Churchill in the vicinity of Gordon Point in the Churchill Wildlife Management Area (see Figure 7.2). To view the bears in relative safety and comfort, visitors board tundra vehicles, which are large-wheeled bus-like vehicles that can accommodate up to 50 people. Two companies currently share licenses to operate 18 vehicles in the Churchill Wildlife Management Area (Tundra Buggy and Great White Bear) and, typically, a polar bear viewing tour takes about seven hours during which tourists can expect to see polar bears as well as arctic fox, ptarmigan and snowy owl. Helicopter viewing of the bears also occurs over both the Churchill Wildlife Management Area and Wapusk National Park. In addition, both polar bear viewing operators own a mobile tundra lodge, allowing visitors to stay on the tundra overnight (Lemelin and Smale 2006). In 2003, the polar bear viewing industry was calculated to have added C$2 million to the Churchill economy (Lemelin 2004). Although no exact visitation data is available, between 6,000 and 10,000 polar bear viewing tourists are estimated to travel to Churchill (Dawson *et al.* 2010).

*Figure 7.2* Polar bear viewing from a tundra vehicle in the Churchill Wildlife Management Area (Photo credit: Jackie Dawson)

## Emergence of a last chance tourism destination

The following section illustrates the emergence of Churchill as a last chance tourism destination by drawing on perspectives from both visitors and local stakeholders outlined in a number of existing studies conducted in Churchill.

Across northern Canada, it is becoming clear that the phenomenon of last chance tourism is more than media hype, as one long-standing Nunavut stakeholder suggested:

> doom tourism – it has been floating around Nunavut for the last 5 years I'd say – people wanting to see the arctic and the critters here in advance of them doing a swan dive due to climate change . . . My sense from the visitors I talk to is that there is a sense of urgency about their visits – not wanting to put visits off, but to see things while they are still authentic . . . The WWF last lonely bear sitting on an ice floe has certainly done a number on the psyche of the southern Canadian population, the USA and Europeans too.

Although last chance tourism is largely denied by the local tourist industry, this type of statement confirms observations that the vulnerability of polar bears in the region is an increasing motivation for tourists to travel to Churchill (Dawson *et al.*, 2010).

During the 2007 polar bear viewing season, 334 tourists were surveyed onsite in Churchill (Dawson *et al.* 2010). The surveys were administered in order to examine both tourists' perception of climate change in general, and the role that climate change played in their decision to travel to Churchill. Results of the 2007 visitor survey revealed that demand for viewing experiences is not likely to decline in the short term. Over 82 per cent of polar bear viewing tourists indicated that even if environmental conditions altered the population dynamics of polar bears so much that they were able 'to view only a quarter of the number of bears they actually saw on a 2007 visit', they would still visit Churchill in the future. The majority of visitors (72 per cent) indicated that if they 'could not view polar bears in Churchill', for example, if the WHB subpopulation was locally extirpated, they 'would be willing to travel somewhere else so they could view polar bears' and 70 percent 'would be willing to pay more to do so' than they did during their 2007 trip. In addition to the main survey, 18 polar bear viewers were interviewed onsite while viewing polar bears, and 11 of those interviewed revealed a heightened motivation to see the species; for example, 'I thought I better come see the bears because the next time I am in this country they will be all gone' (Dawson *et al.* 2010).

Stewart's (2009) study focused on local stakeholders (e.g. educators, managers, and operators) and noted that climate change was emerging as an important issue. Some stakeholders were convinced that 'the bears will always be here', while others thought that the bears may 'come off [the ice] in Arviat'. An educator claimed, 'If climate models are correct the trends suggest there won't be enough ice on Hudson Bay to support polar bears, ringed seals and different birds. No ice means no bears.' More explicitly an operator claimed that 'reports of declining polar bear populations might actually lead to growth in last chance tourism (at least in the short term) because more people will want to see them'. Not surprisingly, there is little evidence to suggest that tour operators in Churchill consider polar bear viewing as a last chance activity, nor do they market it as such. In an online public forum one of the main operators in Churchill officially noted, 'On behalf of our industry I submit that Churchill tourism operators don't participate in this discussion [of last chance tourism]' (Ord, 2010). Regardless, the work by Dawson *et al.* (2010) reveals evidence that some tourists visiting Churchill to view polar bears between the years 2005 and 2009 were indeed motivated to do so because of a perception that the bears are vulnerable to climate change. The fact that this motivation exists means that last chance tourism is occurring and is likely to continue under the constantly changing conditions brought about by global climate change.

## Discussion

Projections of declining polar bear populations in Western Hudson Bay and elsewhere could have significant implications for the polar bear viewing industry. In the short term (i.e. 20–30 years), climate change is likely to cause an increase in the 'waiting period' (i.e. due to longer ice-free periods), which may lengthen the viewing season for tourism business capitalizing on the polar bear viewing market (Dawson *et al.* 2010). Clearly, the economic implications could be significant for the tourism industry, particularly in Churchill, in the short term. However, as Brock (2008: 43) states 'a tourist boom would be good for arts and crafts. But selling more beadwork would hardly make up for an ecosystem in freefall.' Accordingly, because the health of polar bears is likely to decline as a direct result of the longer ice-free periods, the long-term future sustainability of the polar bear viewing industry under projected climate changes is precarious at best (even despite short-term gains). According to Derocher *et al.* (2004: 171) 'exactly how and when polar bears will respond to climate change in different areas is uncertain', so while the timeframe for such change is unclear and while a number of different scenarios for polar bears may exist in the future (increasing rates of cannibalism, aggressive encounters, emaciated polar bears), the following discussion focuses on one possible scenario: what happens to Churchill and other locations when all the 'last chances to see polar bears' have been exhausted, and conversely, how might the wider region respond?

An important theme to emerge from Stewart's (2009) study in Churchill is that northern communities have been, and will continue to be, 'resourceful and inventive in light of change'. Through its long history of 'explosive boom-or-bust cycles [that have] repeatedly swept through the region' (Lemelin 2005: 187) mainly connected to transportation services and military presence, the community of Churchill is arguably well placed to adapt to the loss of the polar bear, and its multi-million dollar polar bear viewing industry. As Churchill stakeholders identified themselves, 'climate change has the potential to change the landscape of business here in Churchill . . . I don't think we'll lose it completely, we'll adapt. It would be devastating if we lost the ice and what-not. No, we won't die, we would adapt.' As residents put it, 'we have no idea really if the bears will stay here or not, it's not something we can help. It would be nice to see the bears stay on the ice, we'd like more time with them, whether that will happen who knows . . . [but] if the bears do go, we'll have to invent something else.' Similar demonstrations of community resiliency and adaptability to the boom-or-bust cycles have also been reported for the Weenusk First Nation at Peawanuck (Lemelin *et al.* 2010; Lemelin and McIntyre 2011) and the Washaho Cree Nation at Fort Severn. In these First Nations, the mixed economy largely based on traditional livelihood has provided them with an ability to withstand global economic fluctuations. Washaho is particularly interested in expanding its current polar bear tourism opportunities through the development of regional partnerships with Nunavut and with the help of tourism operators (Kakekaspan *et al.* 2010).

The longer ice-free season may provide opportunities for Churchill to become a 'gateway' transportation hub in the Canadian Arctic. The concept of the 'Arctic bridge', a shipping route between Churchill, northern Manitoba and Murmansk in northwestern Russia has long been discussed. Both Churchill and Murmansk are summertime sea ice-free seaports and, under warming conditions, it is claimed that shipping times between North America, Europe and Asia could be reduced significantly via these two ports (Mifflin 2007). With an established rail and air network already in place (the airport received $13.4 million in federal funding for improvements in 2010), and a developing road network, this makes Churchill strategically placed to take advantage of the opportunities associated with commercial shipping through the Hudson Bay region, and beyond (in 2007 the Canadian government pledged $8 million for improvements to the Port of Churchill). As one Churchill stakeholder stated in Stewart's (2009) study, 'five years down the road, I suspect shipping from Russia will launch big things for the port, more jobs, more for the railroad, as well as diamond mining'. In confirmation, another stakeholder said, 'tourism isn't going to be such a big thing in the future, shipping is going to be. The time has come for the Port of Churchill. The Northwest Passage is an incredibly important issue for world trade; although it's still a dangerous and challenging environment for ships, we have a 'shoe in' here as an industrial port.' Despite other potentially attractive economic prospects for Churchill, there are indications that those tourism operators who have successfully developed polar bear viewing in the community may follow the bears and relocate farther north. For example, Arviat, the Inuit community 260 km north of Churchill, has witnessed an increase in polar bear activity (Tyrrell 2006) in recent years and is billed to be the 'next' Churchill. Indeed, destination substitution appears to be happening with one operator offering a fly-in tour to Arviat in 2010 (see Steiger, Dawson and Stötter, this volume). Although numbers were not sufficient to run the tour in its inaugural year, in the marketing of the tour, the operator makes a clear link to the altered distribution of polar bears in the region:

> As the Hudson Bay coast freezes up later and later every year, the polar bears are now moving further north to where the ice freezes up earlier. As a result, they are moving away from Churchill and towards Arviat as the polar bears are already adjusting to the later freeze-up of the Churchill area . . . In close partnership with the local Arviat people, the Arctic Kingdom Polar Bear Cabin camp is set in an area known as 'Polar Bear Alley' [which] derives its name from the fact that the local Inuit are seeing increasing numbers of polar bears come here year over year.
>
> (Arctic Kingdom 2011)

Although Arviat's polar bear viewing industry is embryonic, there is indication that 'communities north of Churchill including Arviat and Repulse are uniquely situated to take advantage of bears heading farther north' (P. Scott, personal communication 2010). Indicative of these opportunities Parks Canada plan to develop the interpretive capacity of local staff based in Arviat. Another tour operator in the region suggested that while the repercussions from climate change on polar bears are disconcerting, they will likely follow the bears and move their base of operation northward to Sila Lodge in Ukkusiksalik National Park (see Lemelin and Dyck 2008). The consideration of polar bear viewing opportunities therefore appear critical to the pending Management Plan for Ukkusiksalik National Park.

As in Nunavut, the First Nations of Weenusk, Fort Severn and even Moose Factory have reported an increase in polar bear sightings and encounters in recent years (Dowsley and Wenzel 2008; Lemelin *et al.* 2010). These concerns have prompted some community and tribal leaders to suggest that better polar bear management systems including more active roles for the Cree people are required (Kakekaspan *et al.* 2010; Lemelin *et al.* 2010). In Ontario, there appears to be three competing perspectives limiting the awareness of, and subsequent expansion of, polar bear viewing in this region. The first is public awareness, few people from this province are aware that polar bears are found in this region. The second perspective suggests that that there are many healthy bears, and that the industry should be expanded to promote this opportunity. A recent *National Geographic* video shot in the James Bay region may provide some impetus for this growth. The third perspective pertains to concerns that the environmental quality in this region is deteriorating, and that polar bears could migrate eastward and northward along the Quebec coastline and towards the Belcher Islands in Nunavut, although this possibility is not supported by most biological studies on this particular subpopulation.

## Conclusion

Most would agree that 'there is hardly any other species that has the potential to capture people's attention more effectively than the *king of the Arctic*' (Vongraven 2009: 323), and as such the polar bear has become one of the iconic symbols of climate change (Slocum 2004). Nowhere else is this more obvious than in the Hudson Bay region, where the plight of the polar bear is seen as a 'microcosm of the worldwide circumstance' of the species (Peacock *et al.* 2010: 94). In the context of climate change, this chapter has documented the current scale and scope of polar bear viewing opportunities in the Hudson Bay region, and illustrates how Churchill, in particular, has emerged as a last chance tourism destination. In addition, the chapter discusses the possible regional responses to the transformation of the species. Although Churchill has acted as a 'honey-pot' for polar bear viewing tourism for a number of decades, it is now clear that this status may be transforming, in line with the bears which are at the core of Churchill's nature-based tourism industry. A form of destination substitution may be taking place, whereby in the long term, other locations in the Hudson Bay region could be better placed to offer a different kind of viewing experience (perhaps to one which is more culturally inspired) to that currently offered in Churchill. In this sense we suggest that Hudson Bay is also a *microcosm of the worldwide circumstance* of last chance tourism destinations, and the region can be considered a test case of how destinations address and adapt to the multi-layered complexities of climate-induced change.

A review of the discussions in this chapter suggests that a number of perspectives are at play in the polar bear viewing sector. The first is based on the resiliency of the species to climate change, and the uncertainty surrounding the polar bear's ability to adapt and move northward. Adapting and shifting current tourism activities northward is also being implemented and/or suggested by certain operators; a trend that is likely to continue. Notwithstanding the possibilities presented by last chance tourism in this region, it is almost certain that challenges lie ahead, not least because beyond the immediate short term, the economic base of destinations such as Churchill is arguably vulnerable, but also because Inuit communities in

Nunavut have a different cultural relationship with polar bears (Tyrrell 2006). However, research and media attention suggest that the appeal of this charismatic threatened species will continue to attract tourists to Hudson Bay. The irony with this adaptive rationale across the region is that shifts in tourism supply (and by association tourism demand) may add strain to an environment (both human and ecological) that already is stressed by the effects of climate change. With this in mind, it is crucial for managers and operators to be mindful of economic and cultural sensitivities across the region, as well as being respectful of managing viewing opportunities of a species in demise.

## Acknowledgements

The authors would like to thank Shawn Mueller from Mount Royal University for producing the map used in this chapter, and for helpful comments received from the two reviewers.

## References

ACIA (2004) *Arctic climate impact assessment: Impacts of a warming Arctic.* Cambridge: Cambridge University Press.

Amstrup, S. C., and Gardner, C. (1994) 'Polar bear denning in the Beaufort Sea', *Journal of Wildlife Management* 58: 1–10.

Arctic Kingdom (2011) 'Arviat fly-in polar bear and northern lights cabins'. Online: arctickingdom.com/trips/arviat/ (accessed 15 April 2011).

Brock, P. (2008) 'The town that can't wait for the world to warm up', T*he Sunday Times Magazine*. Online: www.timesonline.co.uk/tol/news/environment/article3362887.ece (accessed 15 April 2011).

Clark, D. (2003) 'Polar bear-human interactions in Canadian national parks, 1986–2000', *Ursus* 14(1): 65–71.

Dawson, J., Stewart, E. J., Lemelin, R. H., and Scott, D. (2010) 'The carbon cost of polar bear viewing tourism in Churchill, Canada', *Journal of Sustainable Tourism* 18(3): 319–336.

Derocher, A. E., Lunn, N. J., and Stirling, I. (2004) 'Polar Bears in a warming climate', *Integrative and Comparative Biology* 44(2): 163–176.

Dowsley, M., and Wenzel, G. (2008) 'The time of the most polar bears: a co-management conflict in Nunavut', *Arctic* 61(2): 177–189.

Environment Canada (2009) 'Backgrounder conservation of polar bears in Canada'. Online: www.ec.gc.ca/default.asp?lang=En&n=714D9AAE-1&news=18E4D45A-CB74–41EE-B1A4-DFCCFF4B8173 (accessed 26 April 2011).

Explore Nunavut (2011) 'Explore the Arctic – Welcome to Nunavut'. Online: www.explorenunavut.com/ (accessed 26 April 2011).

Falkingham, J., Chagnon, R., and McCourt, S. (2001) 'Sea ice in the Canadian Arctic in the 21st century', paper presented at the 16th International Conference on Port and Ocean Engineering under Arctic Conditions (POAC), Ottawa, Ontario, 12–17 August 2001.

Furgal, C., and Prowse, T. D. (2008) 'Northern Canada,' in D. S. Lemmen, F. J. Warrent, J. Lacroix and E. Bush (eds.) *From impacts to adaptation: Canada in a changing climate*, Ottawa: Government of Canada.

IPCC. (2007) *Intergovernmental Panel on Climate Change. Climate Change 2007: The physical science basis: Summary for decision makers*. Paris: IPCC.

Kakekaspan, M., Miles, T., Lemelin, H., Dowsley, M., Beaulieu, M., Taylor, M., *et al.* (2010) *Giving voice to bear: Cree Observations and documentation of Wabusk in Ontario*, Centre for Northern Studies/KORI.

Lemelin, R. H. (2004) 'The integration of human dimensions with the environmental context: A study of polar bear observers in the Churchill Wildlife Management Area, Churchill, Manitoba', unpublished PhD thesis, University of Waterloo, Waterloo, Ontario, Canada.

Lemelin, R. H. (2005) 'Wildlife tourism at the edge of chaos: Complex interactions between people and polar bears in Churchill, Manitoba', in F. Berkes, R. Huebert, H. Fast, M. Manseau and A. Diduck (eds.) *Breaking ice: renewable resources and ocean management in the Canadian Arctic*, Calgary: University of Calgary Press.

Lemelin, R. H., Dawson, J., Stewart, E. J., Maher, P., and Lück, M. (2010a) 'Last chance tourism: boom, doom, and gloom', *Current Issues in Tourism* 13(5): 477–493.

Lemelin, R. H., Dowsley, D., Walmark, B., Siebel, F., Bird, L., Hunter, G., *et al.* (2010b) 'Wabusk of the Omushkegouk: Cree-Polar Bear *(Ursus maritimus)* Interactions in Northern Ontario', *Human Ecology* 38: 803–815.

Lemelin, R.H., McIntyre, N., Koster, R., and Johnston, M. (2010c) 'Climate disruption and the changing dynamics of polar bear – human interactions in Northern Ontario: A case study of polar bear management in Polar Bear Provincial Park, Ontario, Canada', in M. Hall and J. Saarinen (eds.), *Tourism and Change in Polar Regions: Climate, Environments and Experiences*, New York: Routledge.

Lemelin, R. H., and Dyck, M. (2008) 'New Frontiers in marine wildlife tourism: An international overview of polar bear tourism management strategies', in J. E. S. Higham and M. Lück (eds.), *Marine wildlife and tourism management: Insights from the natural and social sciences*, New York: CAB International.

Lemelin, R. H., and McIntyre, N. (2011) 'Resiliency and tourism in Ontario's Far North: The social ecological system of the Weenusk First Nation at Peawanuck', in P. Maher, E. J. Stewart and M. Lück (eds.), *Polar tourism: Human, environmental and governance dimensions*, Elmsford, NY: Cognizant Communication Corporation.

Lemelin, R. H., and Smale, B. (2006) 'Effect of environmental context on the experience of polar bear viewers in Churchill, Manitoba', *Journal of Ecotourism* 5(3): 176–191.

Lunn, N. J., Stirling, I., Andriashek, D., and Kolenosky, G. B. (1997) 'Re-estimating the size of the polar bear population in Western Hudson Bay', *Arctic* 50: 234–240.

Maher, P., and Lemelin, R. H. (2011) 'Northern exposure: Opportunities and challenges for tourism development in Torngat Mountains National Park, Labrador, Canada', *Polar Record* 47(1): 40–45.

Mifflin, M. (2007) 'Arctic sovereignty: A view from the north', *Policy Options* (May): 55–58.

Obbard, M. E., McDonald, T. L., Howe, E. J., Regehr, E. V., and Richardson, E. (2007) 'Polar bear population status in southern Hudson Bay, Canada 1984–2005', Administrative Report, Anchorage: USGS Alaska Science Center.

Ord, C. (2010) 'Last chance tourism? Vanishing destinations and tourism's role in saving them'. Online: www.yourtravelchoice.org/2010/12/last-chance-tourism-vanishing-destinations-and-tourisms-roles-in-saving-them/> (accessed 26 April 2011).

Parkinson, C. L., and Cavalieri, D. J. (2008) 'Arctic sea ice variability and trends, 1979–2006', *Journal of Geophysical Research-Oceans* (113)

Parks Canada. (2007) *Wapusk National Park Management Plan*. Parks Canada. Online: www.pc.gc.ca/pn-np/mb/wapusk/plan.aspx (accessed 26 April 2011).

Parks Canada. (2011) 'Ukkusiksalik National Park of Canada'. Online: www.pc.gc.ca/eng/pn-np/nu/ukkusiksalik/index.aspx (accessed 26 April 2011).

Peacock, E., Derocher, A. E., Lunn, N. J., and Obbard, M. E. (2010) 'Polar Bear ecology and management in Hudson Bay in the face of climate change', in S. H. Ferguson, L. L. Loseto and M. Mallory (eds.), *A little less Arctic: Top predators in the world's largest northern inland sea, Hudson Bay*, Dordrecht: Springer.

Polar Bear Specialist Group (2011) 'Population status reviews'. Online: pbsg.npolar.no/en/status/ (accessed 26 April 2011).

Polar Bear Specialist Group (2011) 'Summary of polar bear population status per 2010'. Online: pbsg.npolar.no/en/status/status-table (accessed 26 April 2011).

Portman, T. (2004) 'Tundra struck: exploring Ontario's polar bear provincial park'. Online: www.cnf.ca/magazine/fall04/tundra.html.

Regehr, E. V., Lunn, N. J., Amstrup, S. C., and Stirling, I. (2007) 'Effects of earlier sea ice breakup on survival and populaiton size of polar bears in western Hudson Bay', *Journal of Wildlife Management* 71: 2673–2683.

Sahanatien, V., and Derocher, A. (2006) 'Changing sea ice-scapes and polar bear habitats in Fox Basin, Nunavut Territory, Canada, (1979–2004)', paper presented at the International Conference on Bear Research and Management.

Schliebe, S., Rode, K. D., Gleason, J. S., Wilder, J., Proffitt, K., Evans, T. J., *et al.* (2008) 'Effects of sea ice extent and food availability on spatial and temporal distribution of polar bears during the fall open-water period in the Southern Beaufort Sea', *Polar Biology* 31(8): 999–1010.

Scott, J. B. T., and Marshall, G. J. (2010) 'A step-change in the date of sea-ice breakup in Western Hudson Bay', *Arctic* 63(2): 155–164.

Slocum, R. (2004) 'Polar bears and energy-efficient lightbulbs: strategies to bring climate change home', *Environment and Planning D: Society and Space* 22: 413–438.

Stewart, E. J. (2009) 'Resident attitudes toward tourism: Community-based cases from Arctic Canada', unpublished thesis, University of Calgary Alberta, Canada.

Stirling, I. (1998) *Polar bears*, Markham: Fitzhenry and Whiteside.

Stirling, I., and Derocher, A. E. (1993) 'Possible impacts of climatic warming on polar bears', *Arctic* 46: 240–245.

Stirling, I., and Parkinson, C. L. (2006) 'Possible effects of climate warming on selected populations of polar bears (Ursus maritimus) in the Canadian Arctic', *Arctic* 59(3): 261–275.

Taylor, M. K., Lee, J., Laake, J., and McLoughlin, P. D. (2006) *Estimating population size of polar bears in Foxe Basin, Nunavut using tetracycline biomarkers*, Iqaluit: Department of Environment, Government of Nunavut.

Tivy, A., Alt, B., Howell, S., Wilson, K., and Yackel, J. (2007) 'Long-range prediction of the shipping season in Hudson Bay: A statistical approach', *Weather and Forecasting* 22: 1063–1075.

Towns, L., Derocher, A. E., Stirling, I., and Lunn, N. J. (2010) 'Changes in land distribution of polar bears in Western Hudson Bay', *Arctic 63*(2): 206–212.

Tyrrell, M. (2006) 'More bears, less bears: Inuit and scientific perceptions of polar bear populations on the west coast of Hudson Bay', *Etudes/Inuit Studies* 30(2): 191–208.

Vongraven, D. (2009) 'The ballyhoo over polar bears', *Polar Research* 28: 23–326.

# Section II

# Media and marketing

# 8 Last chance tourism?

## Public sector views of marketing endangered tourism destinations in North America

*Daniel H. Olsen, Rhonda L. Koster, and Nicki Youroukos*

### Introduction

'Last chance tourism', closely aligned to other similar terms such as 'climate tourism', 'disappearing destinations tourism', 'climate sightseeing', 'doom tourism' or 'doomsday tourism',[1] has slowly become a popular catch-phrase in the media, the private tourism industry, and to a lesser extent tourism academics, to describe the phenomena of vanishing tourism destinations and amenities and the need for people to travel to and consume these destinations before it is too late (Dawson *et al.* 2010; Eijgelaar *et al.* 2010; Lemelin *et al.* 2010). The cause of these vanishing destinations can range from global warming and the accompanying future rise in sea-levels and drought conditions to urbanization and other forms of development to human neglect or the purposeful destruction of natural and human-built environments (Salkin 2007; Ruiz 2008; Frank 2009; Lemelin and Johnson 2008; Smith 2008).

A recent special issue of *Newsweek* (2010) listed 100 places around the world that are endangered due to the potential future effects of global warming. While the editors of the special issue suggest that the purpose is to leave readers with 'an appreciation for the sheer beauty of the earth' and to encourage readers to 'get to work to make sure [these destinations do not] slip away', this list, in essence, gives readers a 'must-visit' list of tourism destinations to visit before rising sea and surface temperatures fundamentally change the nature and quality of these places.

The *Newsweek* list is not unlike that of the list of disappearing destinations provided in the book *Frommer's 500 Places to See Before They Disappear* (Travis 2009). In addition to destinations threatened by global warming, *Frommer's* also includes human-made attractions such as ancient ruins, cities, battlefields and other historical and architectural landmarks, and unique landscapes. In some cases, locations such as Fenway Park in Boston, The Forbidden City in China, the Taj Mahal in India, and the Pyramids of Giza in Egypt are highlighted because they are either being purposefully destroyed (e.g., Fenway Park) or because over-visitation by tourists, urbanization, and pollution have degraded the cultural and built fabric of these sites.

While Dawson *et al.* (2010) argue that it is too soon to label last chance tourism a marketing trend, one could suggest that last chance tourism has great future marketing potential for destinations that are labeled as 'last chance' or 'must-see'. There is slowly growing evidence that last chance tourism is a factor in the tourist decision-making process. For example, Dawson *et al.* (2010) noted in their study of polar bear viewing in Churchill, Manitoba that last chance tourism was a motivating factor for tourists who wanted to see the polar bears in their natural habitat before this was no longer possible. Eijgelaar *et al.*'s (2010) survey of Antarctic travelers suggests that seeing the Antarctic before it is gone was an important motive for travel to this region. MacInnis (2006) notes that for a few years tourists have been rushing to the Rhone Glacier in Switzerland to observe in their words a 'dying glacier'. Nevertheless, there is a need to explore the extent to which last chance tourism is a motivating factor for people to travel to other destinations outside of the Arctic and Antarctic regions as well as destinations that are disappearing for reasons other than climate change.

This chapter focuses on the question of whether or not tourism marketers and stakeholders in destinations labeled as last chance tourism destinations, identified in publications such as *Frommer's* and *Newsweek*, use this designation as a part of their marketing and promotional efforts, and if so, how? As well, what are their opinions about the use of last chance tourism in future marketing and promotional efforts to attract visitors to their destinations? In particular, the focus of this study was on how marketers and managers from the public sector view the phenomenon of last chance tourism. This focus on the public sector was purposeful, in part because the majority of the destinations chosen for this study (see below) fell under the jurisdiction of the public sector (e.g., national parks, cities). Also, the little research conducted on the marketing of last chance tourism has focused on the use of last chance tourism by tourism operators (Lemelin *et al.* 2010) rather than those in the public sphere who manage these destinations. Public sector employee views of last chance tourism is important to consider, because, for example, if last chance tourism is being used by the tourism industry to get tourists to Antarctica (Eijgelaar *et al.* 2010) and the Arctic (Leahy 2010), and if visiting these fragile natural and human-build environments can speed up the 'last chance' element of these destinations (Dawson *et al.* 2010), particularly when transportation emissions contribute to climate change and tourist visitation damages the attributes of these locations (Eijgelaar *et al.* 2010), then those in the public sphere such as protection agency personnel responsible for the management and protection of these fragile destinations will probably have differing perspectives on the use of last chance tourism as a catch-phrase for motivating travelers to visit their destinations.

## Methodology

In order to determine the views of public sector employees regarding last chance tourism as a marketing tool, the authors selected destinations listed in *Frommer's 500 Places to See Before They Disappear* (Travis 2009) and Newsweek's (2010) *100 Places to Remember Before They Disappear*. There were two main criteria for selecting destinations to be included in this study. First, the authors chose destinations located in North America for inclusion, in part because the authors were more familiar with the North American destinations. As well, the managers or marketers of these destinations could be contacted in English. Second, the destinations needed to be described as endangered due to future environmental causes, such as sea and land temperature rise. This specific focus on environmental causes was necessary to ensure consistency between the *Newsweek* and *Frommer's* lists of endangered destinations, for, as noted above, the *Frommer's* list was more inclusive regarding the types of threats a destination faced. The list of destinations chosen for this study is shown in Table 8.1.

Of the 22 destinations in this study, only two were urban destinations, specifically the cities of Chicago, Illinois (threatened by desert-like conditions) and Manhattan, New York (threatened with rising sea levels). The rest of the destinations were national parks and other natural areas, the majority of them found in the United States.

The managers of the natural destinations and the city or regional government tourism officials of cities or larger regions as listed in Table 8.1 were contacted by phone and asked to participate in this study by filling out a short questionnaire that was subsequently emailed to them. The questionnaire focused on four key questions:

1. Are you aware that your destination has been listed in either *Frommer's* and/or *Newsweek*?
2. Would you consider this designation as a 'last chance tourism' destination a benefit to your destination marketing-wise? Why or why not?
3. Is the designation of your destination as a 'last chance tourism' destination used in any of your tourism promoting materials? If so, how? If not, why not?
4. Do you foresee 'last chance tourism' designations as being an important part of future tourism marketing? Why or why not?

In addition, participants were also asked if they were willing to have their names and affiliations noted in the write-up of this study. While the majority of participants were willing to have their identities and affiliations used, in order to fulfill the wishes of those who did not want to be identified in this study the authors have attempted to maintain the anonymity of the participants by referring to them in generic terms throughout this chapter rather than using pseudonyms.

Of the 22 destinations contacted (see Table 8.1), 11 questionnaires were filled out and returned to the authors, giving a response rate of 50 per cent. All of the destinations from which questionnaires were received were nature-based, as questionnaires were not received from any of the urban-based destinations. This lack of response from the urban destinations were in part because when contacted, city officials had a difficult time finding an appropriate and/or willing person to fill out the questionnaire. The responses received by the authors were thematically compiled, coded, categorized and analyzed based upon the themes inferred from the questionnaire questions (e.g., Cope 2010; Dixon 2010)

*Table 8.1* List of North American destinations and/or physical landforms listed as endangered or 'must see' tourist attractions due to climate change by *Frommer's* (Travis 2009) and *Newsweek* (2010).

| *Frommer's Sites* | *Country* | *Protected Area Designation (if any)* | *Setting (i.e. urban-rural)* | *Cause of environmental concern (i.e. sea level rise)* | *Focus (i.e. natural/cultural)* |
|---|---|---|---|---|---|
| Acadia National Park, Maine | USA | National Park | Rural | Sea level rise | Natural |
| Aransas National Wildlife Refuge, Texas | USA | Wildlife Refuge | Rural | Climate change | Natural |
| Bad Branch Preserve, Kentucky | USA | Bad Branch Falls State Nature Preserve | Rural | Climate change | Natural |
| Columbia Ice Fields, Alberta | Canada | Jasper National Park | Rural | Climate change | Natural |
| Death Valley, Nevada | USA | National Park | Rural | Climate change | Natural |
| Denali National Park, Alaska | USA | National Park | Rural | Climate change | Natural |
| Glacier National Park, Montana | USA | National Park | Rural | Climate change | Natural |
| Golden Gate National Recreation Area, California | USA | National Recreation Area | Urban | Climate change | Natural/Cultural |
| Great Sand Dunes National Park, Colorado | USA | National Park | Rural | Climate change | Natural |
| Isle Royale National Park, Michigan | USA | National Park | Rural | Climate change | Natural/Cultural |
| Kenai Fjords, Alaska | USA | National Park | Rural | Sea level rise | Natural |
| Lake Mead, Nevada | USA | None | Rural | Climate change | Natural/Cultural |
| Mesa Verde National Park, Colorado | USA | National Park | Rural | Climate change | Cultural/Natural |
| Redwood Forests, California | USA | The Redwood National and State Parks | Rural | Climate change | Natural |
| Rocky Mountain National Park, Colorado | USA | National Park | Rural | Climate change | Cultural |
| Santa Catalina Island, California | USA | Marine protected area | Marine | Increasing ocean temperatures | Natural |
| The Battery (Southern Manhattan), New York City | USA | Public park | Urban | Climate change | Cultural |
| Big Sur, California | USA | Marine protected area | Marine | Climate change | Natural |
| Charlevoix, Quebec | Canada | UNESCO Biosphere Reserve | Rural | Climate change | Natural |
| Chicago, Illinois | USA | None | Urban | Climate change | Cultural |
| Kauai, Hawaii | USA | Hanalei National Wildlife Refuge<br>Huleia National Wildlife Refuge<br>Kilauea Point National Wildlife Refuge | | Climate change | Cultural |
| Western Hudson Bay | Canada | None | Rural | Climate change | Cultural/natural |

## Results

### *Awareness of last chance tourism*

None of the respondents were aware that their destination was a part of the *Frommer's* or *Newsweek* lists as an endangered or last chance destination. In fact, the majority of them had never heard of such a designation list before, and a couple of respondents were quite concerned with their destination being listed by either *Frommer's* or *Newsweek* as a 'last chance' destination. They asked the authors who had made the lists of endangered destinations and what criteria the editors of *Frommer's* or *Newsweek* had used to create their lists. This concern was, as discussed below, due to the negative connotations that last chance tourism has with their mandates of sustainability and protecting natural resources for future generations. A few of the respondents whose destinations were protected areas suggested that the surrounding private tourism operators or businesses and the local Chamber of Commerce might be more aware of the designation and use it in their marketing strategy.

### *Last chance tourism as a benefit to destinations*

When discussing the potential benefits and drawbacks associated with last chance tourism, one respondent felt that a last chance tourism designation might be beneficial for their destination because 'most visitors here are very concerned about the environment . . . I believe [last chance tourism] would be aimed at the right target audience'. In contrast, however, many of the respondents did not see last chance tourism designations as beneficial to their destinations, as they felt that this designation had the potential to create a negative impression of the destinations in the minds of potential visitors. As one respondent stated, 'I think people should visit this beautiful area to participate in its preservation, not attend its funeral'. Another respondent noted that 'we are not in the doom and gloom trade for the most part', again emphasizing the negative connotations the designation might have on visitors. A different respondent stated that 'the catch phrase "last chance" implies that there is no hope for this dreaded place so come now before it becomes totally inhospitable'.

Other respondents viewed last chance tourism as directly conflicting with their mandates and founding principles to protect the natural area over which they had stewardship, and therefore found last chance tourism as being at odds with sustainable management strategies that were already in place at these destinations. For example, one respondent noted that the 'promotion of any . . . tourism assets as being under threat as a call to action would seem to be incongruous with [sustainable tourism] principles'. Another respondent felt that last chance tourism as a 'catch phrase is short term and not looking to the long term welfare of this place'. Further, another respondent felt that the designation of a last chance tourism destination 'implies surrender, defeatism, and perhaps a relinquishing of responsibility'. They went on to explain that protected area management mandates revolve around research, protection and education of the public, and to assist in managing natural areas for their evolution over time, so that no one ever 'will be facing the last chance' to experience these destinations. As such, last chance tourism, in the minds of many of the respondents, but especially protected area personnel, is a term that is at odds with sustainable management strategies and ideals.

Other respondents saw the designation of last chance tourism as a baseless 'marketing ploy to sell guidebooks' that does not have 'any science-based determination'. While acknowledging that they are preparing for and studying the effects of climate change on the protected area ecosystem, this group of respondents did not see their destination as being under threat of disappearance, but rather evolution and change. Additionally, respondents had differing views about tourism and protected areas. One view was that while protection of the environment is critical, it is also their responsibility to educate and manage visitors to the area and their consequent effects on the ecosystem. Another view, from a more protectionist perspective, was that the destination was not for visitors, but about ecosystem protection and scientific monitoring. One respondent was somewhat neutral in his/her assessment. He/she recognized the fact that a destination's designation as a protected area results in increased visitation and that this then requires management of those visitors to ensure environmental sustainability. Further, he/she acknowledged that climate change is having an impact on protected areas and that it too requires management and provides opportunities for education. In addition, he/she suggested that climate change is a 'sensational enough story on its own without much marketing at all'. However, he/she was concerned that a last chance tourism designation may result in lesser visitation at some point in the future if, for example, 'some icon site or species has changed or disappeared'.

### *Use of last chance tourism in marketing efforts*

Since most respondents were unaware that their sites had been designated a last chance tourism destination, and considering the novely of the term, it was not surprising to find that none of the destinations used last chance tourism in their marketing efforts. In keeping with the dominant opinion that such designations were not beneficial to their destinations, respondents who dealt with the marketing of their destinations indicated that, in addition to the lack of awareness of the moniker 'last chance tourism', they would not use the designation in their marketing for a variety of reasons. For example, it was felt that the use of last chance tourism would be inappropriate in their marketing efforts, as it would give a negative view of the destination to potential visitors, possibly making them feel 'depressed and helpless' regarding the perceived environmental degradation currently occurring at these destinations. In addition, the use of this label may suggest to potential visitors a substandard experience because the site is currently degraded, or soon will be. As one respondent noted, the 'designation may imply to visitors that the attraction or experience has diminished or become degraded from what it was prior to being labeled as a "last chance" destination' and as such its use would in essence be promoting a substandard experience to visitors.

In addition, at many of the destinations the focus was on wilderness protection and research, not consumptive recreational use. In fact, in the US, National Park Service personnel are prohibited from engaging in marketing of their protected areas due to their focus on wilderness management and protection. As such, the use of last chance tourism went against their legislated mandates. As one respondent put it,

> We feel our park protects a fabulous natural and cultural resource which is the property and heritage of all citizens – and that's where we put our energy. The messages we want to convey are clearly laid out in our legal mandate and supporting legislations, and are along the lines of 'these things are here for you and your children.' The concept of 'you'd better come now because your kids won't get to' isn't really part of our culture.

Another respondent commented, 'Our mission . . . is to preserve and protect this unique place for today and *future generations*. A designation like last chance leaves little hope for those future generations.'

### *Last chance tourism in future marketing*

Only two respondents felt that the last chance tourism designation might be part of future marketing strategies. One of the respondents, who stated that their local Chamber of Commerce was responsible for the marketing of their protected area, felt that such a designation would provide an opportunity to reach additional and new visitors. The other respondent felt that 'local businesses will try to capitalize on the last chance designation to pull in more visitor dollars' and that 'some sites, mainly those whose funding comes mainly from entrance and other visitor fees, might try to use the designation, whether it is accurate or not, to increase their visitation and thus their revenues'.

However, the majority of respondents did not think that a last chance tourism designation would be used for future marketing, in part because they felt that the use of last chance tourism as a marketing focus, as noted earlier, would promote their destinations in a negative light – as places that have already become degraded, hence a degraded natural experience. In the minds of many respondents, visitation to protected areas should be about creating stewards so that people will make necessary changes in their lifestyles to protect the natural environment. However, one respondent noted that 'to date, the real push for this type of marketing seems to be driven by the commercial and economical proponents who see a way to make money off people's desire to say they are one of the few who got to see or experience' a particular destination. Thus, while last chance tourism marketing may be important for individual private tourism operators who have attractions that are under threat, these respondents agreed that the moniker was incongruent with the long-term goals of protected areas. As one respondent put it, the goal of protected areas 'does not seem to mix too well with the "collect it now" feeling that "last chance" tourism conveys'. Another respondent suggested that,

> If this type of marketing could truly bring about a greater appreciation of the impacts we have on this environment and really engage people to make changes in their personal behaviors (being more energy efficient and not so wasteful), then we could see some benefit. But as it is now, the only benefit would be short-term gain for a few and a larger impact to the resources we are mandated to protect.

## Discussion

In some ways it was not surprising to find that respondents had not heard that their destinations had been listed as a last chance destination, given the emphasis public sector managers and marketers have on management practices or positive marketing messages based on sustainable principles. Of some interest was the way in which a few respondents were taken aback when they heard that their destination had been listed in either *Frommer's* or *Newsweek* as an endangered or 'last chance' destination. These respondents wanted to talk at length regarding why their destination had been labeled as such, and in a couple of instances this concern almost led them to decline participation in the study. In these instances, links to the *Frommer's* and *Newsweek* publications were provided, and once the respondents were able to look at the publications they agreed to participate. In a sense, sending the respondents who participated in the study the links to these publications became a reward for their participation, in that they gained information that they did not have previously about the way their destination was being represented in the popular literature.

The fact that the respondents had not heard about their destination's last chance tourism designation brings up the question as to the purposes of the authors of *Frommer's* and *Newsweek* for producing these lists and the intended audiences of these lists. In the introduction to *Frommer's*, the author states that he wrote under the assumption that the places in the book were 'on the brink of extinction'. However, the author goes on to explain:

> I discovered more positive developments than I expected. Some of the destinations I've included in this book are already on the road to being saved, usually because they've been championed by preservationists with a will to make a difference. And even in cases where a site has been lost, or irrevocably damaged, it often has become a rallying point for activists, inspiring them to fight on so that the same mistakes aren't made again. Many of the case studies in this book are reasons for hope, not despair, and the more support we can lend them, the better.
>
> (Travis 2009: vi)

The intended purpose of the *Frommer's* book was aimed at giving eco-tourists a list of places to 'enjoy', 'for in the process of cherishing these natural and cultural wonders, we renew our commitment to preserving them' (Travis 2009: vi). This was in essence the same motivation for the editors of the *Newsweek* special issue to create their list of 100 endangered places. According to the online edition of this publication, not only does the content of the interactive website remind readers that 'human beings also occupy a narrow ecological niche', but also that readers should have 'an appreciation for the sheer beauty of the earth' and remember 'what's at stake if we are complacent in the face of rapidly changing climate. Read it, marvel at what we have, and then get to work to make sure it does not slip away.'

In essence, then, the purpose of both publications is to act as a 'call to action' to protect and preserve these endangered destinations. However, these lists are also a 'call to visit' these destinations, so to speak: to see them 'before' or 'in case' they disappear. As such, these lists are well suited to appeal to those travelers who 'collect places' and those who can be considered 'ecological fatalists'. People who collect places are motivated by a number of different factors to visit a destination, such as being able to check off or enumerate a destination from a list, to boast of where they have traveled, and to impress others by their choice of unique or peculiar destinations (Timothy 1998; see King and Prideaux 2010). In this case, these lists of endangered destinations could be used to impress the acquaintances or friends of visitors to these last chance destinations, particularly if these destinations are destined to disappear in the near future. An ecological fatalist, on the other hand, would be attracted to visiting the destinations on these lists because of the inevitability that current and future management practices will not forestall the disappearance of these destinations, and as such these places should be experienced before they are gone (see Milbank 1993).

The views of both the destination managers and the marketers in this study toward last chance tourism were clearly negative in tone. For both groups of respondents, their mandates revolved around the conservation, preservation, and protection of, and education about, these natural resources. The managers in particular viewed visitors to their destinations as people who cared for the natural environment, corresponding closely to an academic view that protected area visitation is thought to be based on people's concern for and wish to access the natural environment (e.g., Eagles and McCool 2002; Dearden and Rollins 2008); as evidenced throughout this book however, there appear to be multiple reasons why people visit natural areas, many of which may not be as altruistic as previously assumed. Certainly researchers examining ecotourism and sustainable tourism (Wheeller 1993; Munt 1994; Ryan *et al.* 2000; Lemelin *et al.* 2008) have criticized the assumption that visitors to such locations are exclusively motivated by environmental, moral or educational aspirations. The view that last chance tourism is a marketing ploy to increase tourist visitation and revenue could be related to the idea that visitors motivated by the chance to see a destination before it disappears are doing so to collect the place or because of their fatalist views of the environment, which are directly in conflict with the goals of sustainability of these agencies. This negative view toward being labeled a last chance destination may also revolve around the fact that the authors of both the *Frommer's* and the *Newsweek* publications did not outline the methodology involved in selecting the destinations included in their lists. In fact, the author of the *Newsweek* publication stated that their list was 'arbitrary, to be sure, and far from all-inclusive'. Thereby supporting the notion that, 'the designation sounds more to me like a marketing ploy to sell guidebooks than any science-based determination'.

Thus, the negative views of the respondents regarding the future use of last chance tourism as a marketing tool was in some ways not surprising, given their negative perspectives on the 'last chance' tourism label. While recognizing that visitation is economically beneficial for local tourism operators and surrounding community businesses, respondents agreed that trying to increase visitation through such marketing ploys threatens what are already vulnerable and threatened species and their ecosystems for what amounts to short-term economic benefit. It would be interesting to see if there is any discussion between the Chambers of Commerce, who in some cases market natural parks and protected areas, and the managers of these destinations when it comes to the way in which those natural destinations are marketed. Based on this research, it would not be surprising to see conflicting views between small tourism enterprises and Chambers of Commerce and those who manage these destinations regarding the use of last chance tourism as a marketing tool.

## Conclusion

This chapter has summarized what can be considered the beginnings of a project revolving around the marketing of last chance tourism. The preliminary study described here focused on the views managers and marketers in the public sector have about last chance tourism as a marketing and promotional tool to endangered destinations. However, there are many areas in which future research could be considered. For example, it would be instructive in a future study to consider the views of the private tourism sector operators at each of the destinations discussed above, to see how their views of last chance tourism side with or differ from the public sector perspective as mentioned in this chapter. Also, future research at an academic level could delve into endangered destinations that are found outside of the polar regions of the world, as to date the use of the term 'last chance tourism' to largely describe destinations that are endangered by global environmental change has been in cold-weather areas such as the Arctic and Antarctica (e.g., Lemelin and Johnson 2008; Dawson *et al.* 2010; Johnston *et al.* this volume; Lamers *et al.* this volume; Lemelin *et al.* 2010). In addition, there is a need to look at how destinations that are endangered by pollution, urbanization, and cultural factors might use the 'last chance' label to enhance the promotion of their destinations. At the same time, it may be too soon to study in an in-depth manner last chance tourism as a marketing tool until there is more tangible evidence of the effect of global warming on these last chance tourism destinations.

This study also highlights the need to examine the construction and use of 'lists' in tourism discourse. The making of a list is a common activity which helps people to organize and prioritize goals and tasks (Schiffrin 1994). Schuster (2002: 4) argues, however, that lists are not just a descriptive, 'neutral catalogue' of organized ideas and goals, but rather is a 'call . . . for action', where either the creator of the list or the person following the list must do or perform something in order to accomplish a task. In the context of tourism, lists can be a powerful motivational tool to encourage people to travel to places they normally would not visit, and as such the study of lists, whether created as a personalized 'bucket list' or a set of places to see or collect before they disappear, 'can provide a source of data for research into a variety of psychological processes' (Schiffrin 1994: 378) related to both the categorization and the decision-making processes of people regarding which destinations they choose to visit and why (e.g., Papatheodorou 2001). In addition, the study of lists can also add a better understanding of how tourism destinations are represented in promotional literature. Lakoff (1987: 50) suggests that categorization is 'not independent of who is doing the categorizing and on what basis'. In addition, Schiffrin (1994: 383) notes that there is 'subjectivity between constructing an experience and categorizing in the world'. As such, a study of the criteria and the power behind the creation of these types of lists can help determine how these lists are developed, by whom they are developed and who chooses the criteria upon which these lists are based, who is the intended audience, and what are the intended results of developing these lists in the first place.

In many ways it may be too soon to look at the marketing implications of last chance tourism. Last chance tourism is based on future possibilities regarding destinations that are considered by some as endangered due to climate change or other ecological and/or human-made concerns. However, for the cases discussed in this study (with the exception of Glacier National Park), there has not been strong visible evidence of climate change and the resultant destruction of natural habitats or ecosystems at its fullest extent. As such, any future research on last chance tourism marketing should be longitudinal, considering the 'before' or 'in case' they disappear connotations of this type of marketing.

## Note

1 While 'doomsday tourism' in this context refers to the disappearance of destinations, 'doomsday tourism' can potentially also be based on 'end of the world' scenarios, such as those surrounding the year 2000 (Olsen and Timothy 1999) and tourism based on the movie *2012* and the predicted end of the world by the Mayan calendar (e.g., Sanders 2009).

## References

Cope, M. (2010) 'Coding qualitative data', in I. Hay (ed.) *Qualitative Research Methods in Geography*, Oxford and New York: Oxford University Press, pp. 281–294.

Dawson, J., Stewart, E. J., Lemelin, H. and Scott, D. (2010) 'The carbon cost of polar bear viewing tourism in Churchill, Canada', *Journal of Sustainable Tourism* 18(3): 319–336.

Dearden, P. and R. Rollins (eds.) (2008) *Parks and Protected Areas in Canada: Planning and Management*, 3rd edn, New York: Oxford University Press.

Dixon, D. P. (2010) 'Analyzing Meaning' in B. Gomez and J. P Jones III (eds.) *Research Methods in Geography*, Chichester, UK: Wiley-Blackwell, pp. 302–407.

Eagles, P. F. J. and McCool, S. F. (2002) 'Tourism in national parks and protected areas', *Planning and Management*, New York, NY: CABI Publishing.

Eijgelaar, E., Thaper, C. and Peeters, P. (2010) 'Antarctic cruise tourism: the paradoxes of ambassadorship, "last chance tourism" and GHG emissions', *Journal of Sustainable Tourism* 18: 337–54.

Frank, P. J. (2009) 'Travel's most endangered destinations'. Online: www.msnbc.msn.com/id/25737162/ (accessed 2 March 2011).

King, L. M. and Prideaux, B. (2010) 'Special interest tourists collecting places and destinations: a case study of Australian world heritage', *Journal of Vacation Marketing* 16(3): 235–247.

Lakoff, G. (1987) *Women, Fire, and Dangerous Things: What Categories Reveal About the Mind*, Chicago/London: The University of Chicago Press.

Leahy, S. (2010) 'Tourism at the end of the world'. Online: www.tierramerica.info/nota.php?lang=eng&idnews=1970 (accessed 14 June 2010).

Lemelin, R. H., Dawson, J., Stewart, E. L., Maher, P. and Lueck, M. (2010) 'Last chance tourism: the boom, doom, and gloom of visiting vanishing destinations', *Current Issues in Tourism* 13(5): 477–493.

Lemelin, R. H. and Johnson, M. (2008) 'Northern protected areas and parks', in P. Dearden and R. Rollins (eds.) *Parks and Protected Areas in Canada: Planning and Management*, 3rd edn, New York: Oxford University Press, 294–313.

Lemelin, R. H., Smale, B., and Fennell, D. (2008). 'Polar bear viewers as deep ecotourists: how specialized are they?' *Journal of Sustainable Tourism* 16(1): 42–62.

MacInnis, L. (2006) 'Dying glaciers draw curious to Swiss alpine peaks'. Online: www.planetark.com/dailynewsstory.cfm/newsid/38141/story.htm (accessed 15 June 2010).

Milbank, J. (1993) 'Out of the greenhouse', *New Blackfriars* 74(867): 4–14.

Munt, I. (1994) 'Eco-tourism or ego-tourism?', *Race & Class* 36 (1): 49–60.

*Newsweek* (2010) '100 places to remember before they disappear'. Online: www.newsweek.com/feature/2010/100-places-to-remember.html (accessed 11 June 2010).

Olsen, D. H. and Timothy, D. J. (1999) 'Tourism 2000: selling the millennium', *Tourism Management* 20(4): 389–392.

Papatheodorou, A. (2001) 'Why people travel to different places', *Annals of Tourism Research* 28(1): 164–179.

Ryan, C., Hughes, K. and Chirgwin, S. (2000) 'The gaze, spectacle and ecotourism', *Annals of Tourism Research* 27(1): 148–163.

Ruiz, R. (2008) 'Endangered natural wonders worth seeing'. Online: www.forbes.com/2008/09/15/travel-endangered-ecosystems-forbeslife-cx_rr_0915travel.html (accessed 2 March 2011).

Salkin, A. (2007) '"Tourism of doom" on Tise'. Online: http://www. nytimes.com/2007/12/16/world/americas/16iht-tourism.1.8762449.html?_r=3 (accessed 2 March 2010).

Sanders, J. (2009) 'How Tikal hotels are preparing for the end of the world'. Online: www.jaunted.com/story/2009/12/7/13586/5436/travel/How+Tikal+Hotels+Are+ Preparing+for+the+End+of+the+World (accessed 11 June 2010).

Schiffrin, D. (1994) 'Making a list', *Discourse Processes* 17(3): 377–406.

Schuster, M. J. (2002) 'Making a list and checking it twice: the list as a tool of historic preservation', paper presented at the Association for Cultural Economics International Conference, Rotterdan, The Netherlands. Online: http://harrisschool. uchicago.edu/about/publications/working-papers/pdf/wp_03_03. pdf (accessed 10 March 2011).

Smith, J. (2008) 'Endangered destinations: places, like species, can vanish forever. A look at some unique, imperiled treasures'. Online: www.usnews.com/science/ articles/2008/05/15/endangered-destinations (accessed 2 March 2011).

Timothy, D. J. (1998) 'Collecting places: geodetic lines in tourist space', *Journal of Travel & Tourism Marketing* 7(4): 123–129.

Travis, W. (ed.) (2009) *Frommer's 500 Places to See Before They Disappear*, Hoboken, NJ: Wiley Publishing, Inc.

Wheeller, B. (1993) 'Sustaining the Ego', *Journal of Sustainable Tourism* 1(2): 121–129.

# 9 Advertising World Heritage Sites

## Tour operators and last chance destinations

*Elspeth Frew*

### Introduction

Since tourism is largely dependent on natural resources, namely beaches, coastlines, mountains, forests, lakes, oceans and scenery, these elements are central to the attraction potential of most destinations, with most tourism based on stable and favourable environmental conditions (Gossling and Hall 2006). In the early twenty-first century, the greater populous became more aware of the impacts of climate change, possibly influenced by high-profile, accessible forms of media such as Al Gore's 2006 movie *An Inconvenient Truth* which included dramatic time-lapse photography, images of polar bears stranded on melting ice, and Mount Kilimanjaro denuded of ice and snow.

In that period, media sources published articles on the impact of climate change on the world's best loved natural phenomena, with some writers encouraging tourists to visit various natural sites "before they disappear" (see, for example, Booth 2007; McKay 2007; Smight 2008). The editor of a travel agents' magazine coined the phrase "environmental tourists of doom" to describe tourists who deliberately visit destinations with natural attractions as they believe they will soon disappear due to changes brought about by climate change (Salkin 2007). Popular destinations for these tourists include glaciers, snow fields, rainforests and regions such as the Arctic, Antarctica, Mount Kilimanjaro, and the Amazon, with some travel agents reporting that clients requested trips to see the melting glaciers of Patagonia, the threatened coral of the Great Barrier Reef, and the eroding atolls of the Maldives (Salkin 2007). Thus, the media is replete with articles on famous natural attractions that are already changing significantly, and these writers appear to encourage individuals to visit these natural destinations before they "melt, crumble, fade or drift away, never to be seen or experienced in quite the same way again" (Kendle 2008).

This chapter investigates tour operators' online promotion of tours to a sample of World Heritage Sites which are reported as being negatively affected by climate change. The aim of the study is to determine the propensity among tour operators to use aspects of global warming and climate change to promote their tours. This is particularly illuminating given the highly competitive environment in which tour operators' work, where they are under pressure to maximise the use of the natural resources before they deteriorate beyond use or disappear completely.

## The ecotourism paradox

Ecotourism is defined as a "sustainable, non-invasive form of nature–based tourism that focuses primarily on learning about nature first-hand, and which is ethically managed to be low-impact, non-consumptive, and locally oriented" (Fennell 2008: 24). Ecotourism should incorporate three elements, first, it should involve a nature-based element in which its attractions are primarily based in the natural environment; second, there should be an element of education, learning or appreciation between the ecotourist and the attraction; and, finally, ecotourism should be environmentally and socio-culturally sustainable (Cousins 2007), in a way which "enhances the natural and cultural resource base of the destination and promotes the viability of the operation" (Weaver 2001: 15). Thus, visitation to regions affected by climate change could be described as a form of ecotourism, particularly as the individual tourist learns about the impact of climate change and may be given the opportunity to contribute financially and/or, in kind, to help conserve the region.

Wheeler (2003) notes that this type of tourism is being "ecstatically embraced with unbridled enthusiasm by virtually all sections of tourism" (p. 228) but suggests that this wholehearted and unquestioning acceptance of such forms of tourism is "desperately disturbing" (p. 228) as it conveniently appeases any guilt felt on the part of the tourist, while simultaneously providing increased holiday options and experiences. Although ecotourism has been identified as a form of sustainable tourism expected to contribute to both conservation and development, Tsaur, Lin and Lin (2006: 651) note that "because of inadequate environmental assessments and management, many ecotourism destinations tend to be both hazardous and self destructive" (p. 651). Tsaur *et al.* (2006) also note that many destinations are suffering from the phenomenon of "honoring the name of ecotourism on the surface but destroying the environment in reality" (p. 651) where the overall aim is to attract more tourists. What is not considered is whether positive interactions with local residents can be established, whether the ecological carrying capacities may be exceeded, or whether sustainability can be maintained. Indeed, Weaver (2005) argues that contemporary ecotourism is largely incapable of fulfilling its potential to achieve meaningful environmental and sociocultural sustainability. As a result of such criticism, industry codes such as the Code of Ethics and Guidelines for Sustainable Tourism (Tourism Industry Association of Canada), Environmental Codes of Conduct for Tourism (United Nations Environment Program), Sustainable Tourism Principles (Worldwide Fund for Nature and Tourism Concern), the APEC/PATA Code for Sustainable Tourism and a range of other evaluation programs, certifications and awards were established to encourage environmentally or socially sustainable tourism businesses (Weaver and Lawton 2006).

McCarthy (2007) suggests that visiting destinations before they disappear due to global warming is a logical extension of the growth of ecotourism, particularly because he believes that tourism is broadening out more thematically and is becoming "issue tourism". However, Wheeler (1994) suggests that any form of ecotourism "soothes consciences, demands no sacrifices and allows extended holiday choice while providing an ideal shield, doubling as a marketing ploy, for the tourism industry" (p. 653). Indeed, Kendle (2008) notes that visits to destinations affected by global warming reflects human selfishness to want to see something that others might not be able to in the future. The paradox of this type of ecotourism is that these ecotourists are contributing to global warming by travelling long distances and creating greenhouse gas emissions (Dawson *et al.* 2010). Ironically, the impact of global warming has created some new ecotourism sites. For example, melting glaciers in Greenland revealed islands and underwater sites to explore, that otherwise would have remained hidden (McKay 2007).

## Tour operators and the inclusive holiday industry

Tour operators are the key organisations in the inclusive holiday industry, providing the logistical skills to package travel and arrange accommodation and destination activities in ways which appeal to the travelling public (Laws 1997). Tour operators organise, or package together different elements of the tourism experience and offer them for sale to the public through the medium of a brochure, leaflet or advertisement. A tour operator's package requires at least two elements to be offered for sale at an inclusive sales price and involves a stay of more than 24 hours in overnight accommodation (Page and Connell 2006). Thus, to join an organisation such as the National Tour Association (NTA) as a tour operator, the individual or company must engage in the business of conducting or arranging multiple component travel that is promoted and sold for an inclusive price on a for-profit basis (NTA 2008). Some tour operators specialise in a particular destination or type of holiday and as such are providing services for one, or a few, distinct segments or niches in the market. These small, specialist tour operators offer selected opportunities for individuals or groups to visit destinations or engage in itineraries of particular interest to them; for example, sports, bird watching. Others specialise in adventure holidays perhaps offering a selection of four-wheel drive desert crossings, mountain trekking or guided tours of remote areas. The length and timing of each tour will reflect conditions at the destinations and each tour will be designed to maximise particular interests. These smaller tour operators have a competitive advantage compared to larger tour operators in that they can respond quickly to changing demand, producing new types of holidays for quite small numbers of clients (Laws 1997). These specialist tour operators are often members of associations established to cater to their needs, such as the Association of Independent Tour Operators.

Baloglu and Mangaloglu (2001) explain that tour operators and travel agents have multiple and critical functions in destination marketing efforts; namely, they provide information to potential travellers and develop and promote destination packages. Tour operators have a strong influence on international flows from main generating markets to various destinations and the success of many destinations depends on whether foreign tour operators include them in their programmes (Cavlek 2002). Laws (1997) notes that the key to a tour operator's success is to understand what the customers want, and to provide what they want in convincing and enticing ways. Thus, tour operators compete with each other in terms of both the product they sell and the travel images they create, with these images continually being redefined (Reimer 1990).

Growing numbers of consumers are interested in selecting tours where the operator demonstrates sustainable tourism practices. Information about a tour operator's environmental sustainability helps increase loyalty by tourists towards a particular tour operator. As such this may encourage more tour operators and the tour industry to develop responsible behaviour – if only to secure more business. Thus, the consumer has strong potential to shape the future development of the tourism industry (Milller 2001). Similarly, Campu and Yague (2008) found that the key to tour operators' obtaining loyal consumers is to offer products of high perceived quality. These high-quality trips increase the probability that the consumer will be satisfied, will travel again with the particular tour operator, and will recommend the brand to friends and family.

Several researchers have investigated the use of the internet by tourism operators (see, for example, Cano and Prentice 1998; Wan 2002) but fewer studies have considered tour operators in particular and their websites (see, for example, Rachman and Ritchins 1997; Cai *et al.* 2004; Stepchenkova and Morrison 2006). In their study of US tour operator websites Cai *et al.* (2004) found that a few of the features of the websites were targeted at presenting a trustworthy corporate image, such as customer testimonials, membership of professional organisations, and media articles about the company. They found that one tour operator presented information on its involvement in a wilderness conservation project in China. They note that the inclusion of these aspects on a website creates a powerful marketing tool, as customers may be more inclined to purchase products and services from socially responsible companies.

## Methodology

Study sites were chosen by identifying natural phenomena important to tourism businesses that are being affected by climate change. To do so reference was made to the UNESCO report on World Heritage Sites (2007) and also to the Intergovernmental Panel on Climate Change (IPCC) Working Group II report (2007). The report highlighted the sensitivity, adaptive capacity and vulnerability of natural and human systems to climate change, and the potential consequences of climate change. The report divided the world into the regions of Africa, Asia, Australia and New Zealand, Europe, Latin America, North America, the Polar Regions and Small Islands to describe climate change impacts, vulnerability and adaptability in those areas. For the purposes of this study, each of the chapters from the IPCC (2007) report were examined to determine which natural phenomenon were reported as beginning to be affected by climate change. Table 9.1 summarises the natural phenomenon affected by climate change by world region and reveals there are similar impacts being experienced around the world due to warmer air and sea temperatures, associated melting of snow and glaciers, and rising sea levels.

*Table 9.1* Natural phenomenon affected by climate change by world region

| *World Region* | *Natural phenomenon identified as being affected by climate change* |
|---|---|
| Africa | • Mount Kilimanjaro<br>• mammal species in national parks<br>• wildlife areas and parks<br>• mangroves<br>• coral reefs |
| Asia | • Himalayan glaciers<br>• ski areas<br>• beaches |
| Australia and New Zealand | • Great Barrier Reef<br>• Queensland Wet Tropics<br>• Kakadu Wetlands<br>• alpine areas |
| Europe | • Alps<br>• glaciers |
| Latin America | • Amazonia<br>• tropical rain forests |
| North America | • Great Lakes<br>• glaciers<br>• ski areas |
| Polar Regions | • polar bears<br>• seals<br>• migratory sea birds |
| Small Islands | • beaches<br>• coral reef<br>• Barbados, Bonaire, Kiribati, Tuvalu, Marshall Islands, Maldives |

Source: adapted from IPCC, 2007

This exercise identified the following three World Heritage Sites as being particularly affected by climate change:

- the Great Barrier Reef, in Australia (IPCC name = Great Barrier Reef);
- the Ilulissat Icefjord, Greenland, Denmark (IPCC name = Polar regions);
- the Glaciers in the Kilimanjaro National Park in the United Republic of Tanzania (IPCC name = Mount Kilimanjaro).

A list was then generated to include, large and small, international and domestic tour operators who operate to and within these three natural areas. A variety of sources were used to identify operators including analysis of international, national and regional online tour operator membership directories such as: United States Tour Operators Association (USTOA), Association of Independent Tour Operators (AITO), National Tour Association (NTA), Ecotourism Australia (ETA), Tanzania *Association* of *Tour Operators (TATO), Association of Arctic Expedition Cruise Operators (AECO).* To identify the individual tour operators operating to the three regions, the country name was used to search the membership direc-

*Table 9.2* Members of tour operator associations operating in selected regions in 2008

| | *Great Barrier Reef* | *Greenland* | *Mount Kilimanjaro* |
|---|---|---|---|
| USTOA | 59 (for Australia) | 3 | 2 (for Tanzania) |
| AITO | 14 | 3 | 11 |
| NTA | 23 | 0 | 1 |
| ETA | 31 | n/a | n/a |
| TATO | n/a | n/a | 104 |
| AECO | n/a | 13 | n/a |
| Total | 127 | 19 | 118 |

Key: USTOA – United States Tour Operators Association; AITO – Association of Independent Tour Operators; NTA – National Tour Association; ETA – Ecotourism Australia; TATO – Tanzania Association of Tour Operators; AECO – Association of Arctic Expedition Cruise Operators

tory (i.e., Australia, Tanzania, Greenland) and/or by selecting the most appropriate holiday themes appropriate to each of the three regions, that is, ecotours, wildlife, nature, adventure (for all three regions), cruises (for the Great Barrier Reef and Greenland), safari (for Kilimanjaro), trekking or climbing (for Kilimanjaro), and snorkeling (for the Great Barrier Reef). Table 9.2 summarises the number of tour operators listed by each tour operator association who operate in these areas and had associated websites in 2008.

Recognising that not all tour operators are members of associations, reference was also made to tour operators operating in these three areas who use promotional websites. To this end, the search engine Google was used in July 2008 to identify additional operators. Some tour operators are neither members of associations or have sponsored links. Thus, Google was again used to search for tour operators using the additional terms, "global warming" for Kilimanjaro and Greenland and, "climate change" for Great Barrier Reef. Some tour operators were duplicated between the tour operator association member directories, the sponsored links and the non-sponsored links, so the duplicates were removed. Table 9.3 summarises the number of tour operators operating in each of the three regions.

The websites of the identified 336 tour operators operating in the three regions were examined to determine which operators mentioned aspects of global

*Table 9.3* Number of tour operators operating in the three regions in 2008

| *Natural regions* | *Membership directories of trade associations* | *Google sponsored links* | *Google non-sponsored links* | *Total number of websites examined* | *Number of websites mentioning global warming* |
|---|---|---|---|---|---|
| Great Barrier Reef | 127 | 26 | 5 | 158 | 9 |
| Greenland | 19 | 10 | 1 | 30 | 6 |
| Kilimanjaro | 118 | 25 | 5 | 148 | 15 |
| | | | Total | 336 | 28 |

warming. There were 28 tour operator websites identified that use the terms "global warming" and/or "climate change" on their websites. The websites of those 28 tour operators were examined in more detail using content analysis. Content analysis is a research technique for "making replicable and valid inferences from texts (or other meaningful matter) to their context of their use" (Krippendorff 2003: 18) and is frequently used in tourism to discover how destination images are formed (see, for example, Dann 1988: Lee *et al.* 2006; Markwick 2001; Stepchenkova and Morrison 2006; Xiao and Smith 2006). Similar to the study by Echtner (2002) and Echtner and Prasad (2003), for the content analysis of the written word, the units of analysis were the tour operators' web pages. These were examined for the frequency of descriptor words (verbs, adjectives and adverbs) used in relation to the advertising of the tour operators products. The descriptor words of the verbs, adverbs and adjectives were examined as they convey the means by which the tour was promoted; for example, the "disappearing glacier" or "coral bleaching". Then the key descriptor words relating to the advertising of the tour were highlighted. The text data from each of the tour operator websites identified as mentioning global warming and/or climate change were imported into the software package 'QSR NVivo 8.0' for further content analysis. For ease of discussion, the tour operators were named as follows: GBR1–9 (for operators in the Great Barrier Reef); GRE 1–6 (for operators in Greenland); and, MKJ1–13 (for operators who operate climbs to Mount Kilimanjaro). The web pages of these 28 tour operators were again visited in early 2011 and comparisons were made with the materials downloaded from the same companies in 2008.

As the purpose of this study was to investigate the prevalence among tour operators who use aspects of global warming and climate change to promote their tours, the websites were analysed by the level of coverage. This was analysed in both 2008 and 2011 using the following categories:

- *minor coverage* – one to three sentences mentioning aspects of global warming;
- *moderate coverage* – from four sentences to a couple of paragraphs dealing with aspects of global warming;
- *major coverage* – full page coverage of aspects of global warming, either in the form of a company media release or, the promotion of a tour designed to investigate global warming first-hand in the natural region.

## Results and discussion

### *Minor coverage in 2008*

Five of the tour operator websites used terms such as "believed", "probably", "may be" and "if" in relation to global warming, suggesting they (and perhaps the potential tourist) remain doubtful about the impact of global warming on the natural resources. Two tour operators used the term "sadly" to recognise that changes have now begun to be experienced at the natural destination (Table 9.4).

*Table 9.4* Wording used by minor coverage tour operators

- "These glaciers . . . are, as a result of global warming, **believed** to be receding" (KMJ4).
- "While the glaciers have recently receded, **probably** due to global warming, they remain for now an overpowering and unforgettable sight" (KMJ3).
- "The ice cap of Mount Kilimanjaro . . . **may be** gone in less than 20 years because of global warming" (KMJ13).
- "Global warming **may be** putting pressure on Australia's iconic Great Barrier Reef" (GBR8).
- "**If** global warming continues at its current rate, Greenland's ice sheet could begin to melt this century" (GRE6).
- "It is **believed** that Kilimanjaro . . . is loosing glaciers due to global warming" (KMJ1).
- "**Sadly**, Kilimanjaro is now a prime example of global warming" (KMJ5).
- "**Sadly,** global warming is melting this away . . . in a couple of years there will be very little left" (KMJ12).

To encourage potential tourists to climb Mount Kilimanjaro, the tour operator KMJ12 stated the following: "Possibly for those interested in climbing this gigantic beauty, it is well recommended that you go earlier than later." Similarly, GRE1 in bold letters at the end of an itinerary to Greenland stated "ATTENTION! Due to global warming the sea-ice is getting thinner and thinner every year. Within a few years, these expeditions will no longer be possible. This is the case already with our Svalbard winter-programs." Thus, the minor coverage by these tour operators reflect three aspects – namely, uncertainty about the impact of global warming; some regret about global warming; and a "call to action", that is, encouraging the tourists to visit before the natural phenomena disappear.

### *Minor coverage in 2011*

Of the 11 tour operators, only one tour operator had made any changes to their web page (KMJ13) by removing the section entitled, "The Meaning of Kilimanjaro". This section referred to aspects of global warming and the following sentence had been deleted, "The ice cap of Mount Kilimanjaro is one of the most famous landmarks of Africa, but it may be gone in less than 20 years because of global warming."

### *Moderate coverage in 2008*

The tour operators with moderate coverage of the impact of climate change on the natural regions addressed it in several ways. For example, the website of GBR4 highlighted that a founding member of the company had participated in conservation movements around the world and was a pioneer in ecotourism, which implied their expertise and knowledge in all aspects of the environment. Similarly, the website of KMJ6 noted that their tour leader has recently provided advice to the

National Parks on environmental impact reduction and provided a link to the detailed technical report to which the company had contributed to in 2006 entitled, "A Preliminary Assessment of the viability of siting an Ecological Research Station in the Northern Kilimanjaro Area". This implied that if the tourist purchased the trips organised by these tour operators they would be assured of experiencing the tour with knowledgeable guides.

In a section titled "Let's Talk about Kilimanjaro", one of the Mount Kilimanjaro tour operators (KMJ7) provided information about climbing the mountain under headings such as Physical Fitness, Altitude and Acclimatisation, Personal First Aid Kit, and Photography, etc. In the section entitled "Glaciers and Glaciology", the tour operator discussed in detail how glaciers are formed, and in that context discussed glacial recession, noting that, "if the present rate of glacial recession continues the majority of the glaciers on Kilimanjaro could vanish altogether in the next 50 years" (KMJ7). Similarly, three of the Great Barrier Reef tour operator websites (GBR6, GBR7 and GBR9) provided detailed facts about the ecology of the Great Barrier Reef. In the context of discussing the Great Barrier Reef as a living organism, mention was made of coral bleaching and its associated relationship with climate change. For example, the website of GBR9 stated that "prolonged bleaching can lead to coral death – one more way global warming and periodic climate shifts such as El Nino can change the face of the earth" (GBR9). However, each of the three Great Barrier Reef tour operators implied that it is not only the Great Barrier Reef that is beginning to be impacted by climate change but other reef areas are also facing similar issues. For example, tour operators GBR6 and GBR7 noted the following: "Scientists fear that global warming will continue to cause bleaching on the reefs throughout the world" (GBR6), and "This phenomenon is not exclusive to Queensland's Great Barrier Reef, but has been observed on reefs throughout the world" (GBR7). This suggests that the tour operators are keen to emphasise that the Great Barrier Reef should continue to be visited since all reefs are facing the same issues. Thus, the moderate coverage tour operator websites introduced aspects of global warming within a discussion of the environmental aspects of the tour. Only one tour operator (KMJ9) used scare tactics to promote their tour in a section named "General Information" about Mount Kilimanjaro, when they stated, "If you have not planned to visit Tanzania and climb Mount Kilimanjaro, plan it now! You need to witness the climatic change and its effects. Don't wait to be told or just read as history from books" (KMJ9).

### *Moderate coverage in 2011*

Of the eight tour operators with moderate coverage of climate change in 2008, only one tour operator (GBR3) made any modification to the web page by removing the following sentences, "The Great Barrier Reef is in much better shape than most other reef systems of the world. Climate change is beginning to affect the Great Barrier Reef, as are other human aspects such as nutrient-run off and loss of fish populations." The following updated information relating to climate and the Great Barrier Reef and link was, however, provided to the updated web page: "Visit UNESCO World Heritage Centre for more information."

### *Major coverage in 2008*

Two of the tour operators on the Great Barrier Reef (GBR1 and GBR2) and one operator on Mount Kilimanjaro (KMJ2) provided detailed coverage of the impact of climate change on the natural region but do so in a matter of fact, non-sensational manner. Similarly, four of the tour operators in Greenland (GRE2, GRE3, GRE4 and GRE5) gave major coverage to aspects of global warming. Two of the major coverage tour operators developed climate protection initiatives for tourists. For example, on their website, GRE2 provided a media release page to highlight their programme called the "Polar Ambassador Program" which provides post-expedition support through an online community for participants who want to actively reduce their carbon footprint. Tour operator KMJ11 developed a tour of Mount Kilimanjaro for a maximum of 12 people entitled the "Climate Change Challenge, Mount Kilimanjaro Mission". To be part of the climb, tourists are asked to donate a minimum of $US1,200 per person to a non-profit conservation organisation (plus pay the usual price of $US7,000 to participate in the tour). The tour operator notes that the money donated will be used to buy weather equipment for researchers on Mount Kilimanjaro. Tour operator KMJ11 appears genuine in attempting to be a sustainable tourism operator, noting on their web page that in 2002 they won the World Travel and Tourism Council Green Award for Corporate Social Leadership in Travel and Tourism and, in the same year, they won the Eco-tourism Society of Kenya award for "Demonstrating commitment to a good environmental and socio-economic practices in running a lodge." However, GRE2 appears more commercial in their development of climate protection initiatives, as they conducted a survey of their visitors to assess their opinions regarding global warming and responsible tourism and found that, as a result of their expedition, 60 per cent of the tourists were motivated to protect environmentally sensitive areas. Thus, by providing the tourists with opportunities to be involved in an online community, the tour operator has the opportunity to regularly promote upcoming tours and other products to this captive and interested audience.

In 2007, tour operator GRE4 developed a new tour to "Warming Island", Greenland which is an island recently discovered due to the melting of the ice sheet. The tour in September 2008 was, not surprisingly, listed as being sold out, particularly given that the tour operator notes that the island has been receiving "global attention from major newspapers worldwide, including the New York Times". At the bottom of the webpage there is a link to *The New York Times* article entitled "As Ice Recedes, Interest Surges". Similarly, tour operator GR5 also developed a new tour called "Jewels of the Arctic" which they describe as their "Inaugural Climate Change Voyage" where "special guests open our eyes to the dynamic change that global warming brings to the Arctic". The tour operator invites tourists to "see first hand, as our guest lecturers explain how life above the Arctic Circle is adapting to its surrounding environments as they succumb to the effects of global warming" (GRE5). Similarly, tour operator GRE3 provided a media release page explaining that the tour operator had been involved in a recent expedition to the Arctic entitled "Arctic Expedition for Climate Action", which aimed to highlight changes to the region due to climate warming. Thus, these major coverage tour operators took advantage of the surge in interest in the impact of global warming by providing services to meet demand from serious, committed environmental tourists with higher disposable incomes. This reflects the ability of tour operators to respond to changing demand by quickly developing these global warming-related tours.

### *Major coverage in 2011*

Of the seven tour operators with major coverage of aspects of global warming studied in 2008, one tour operator no longer operates (GBR 2). In 2011, one tour operator (KMJ2) no longer included an article on their web page on the impact of climate change on Kilimanjaro via their "Kilimanjaro and Africa News" page. In 2008 the news page contained an article entitled "Vanishing into Thin Air: The Fate of Kilimanjaro's Glaciers?" which described the disappearance of the glaciers by 2050. Thus, the demand for "last chance tourism" appears to have waned slightly among ecotourists and tour operators perhaps implying apathy, complacency or detachment from this global, complex phenomenon. Whatever the reason for such a decline, there still remains the potential to develop this type of tourism in the future.

The web pages of two of the operators (GBR1 and GRE3) were unchanged from 2008 and, in 2011 provided extensive coverage about the effects of climate change on the Great Barrier Reef (GBR1) with tour operator GRE2 providing a "Summary of the Arctic Expedition for Climate Action". However, the report on the "Arctic Expedition for Climate Action" was published in July 2008 and has not been updated since then, suggesting that the tour ran only in 2008 and not since.

As mentioned earlier, one of the tour operators to Greenland (GRE4) in 2008 promoted a tour called "Voyage of Discovery to Warming Island, Greenland". In 2011, tour operator GRE4 operated tours to the same region but no longer mentions Warming Island in the tour title. Instead the tour is called "Explore our Arctic Wonderland Spitsbergen!" And the tour operator states that, "This voyage offers an ideal opportunity to explore the Arctic and to better understand global warming and future Arctic issues" but provides no further detail on what these Arctic issues may be. However, on the web page the tour operator notes that the company tries to develop new trips every year "to capture the interest of our travelers" and in 2011 a new trip was developed to allow customers to travel with members of the Planetary Society and other "like-minded travelers and stimulating scientific experts who share a love of learning, discovery, and adventure". As such, it may be that customers are currently less interested in "last chance tourism" but remain interested in seeking educational and adventure in remote and pristine environments.

One major coverage tour operator (GRE5) developed an "inaugural Climate Change voyage" in 2008, where "special guest lecturers open our eyes to the dynamic change that global warming brings to the Arctic . . . altered habitat of polar bears and walrus, to diminishing pack ice and receding glaciers". The same tour will operate in 2011 but the term "Climate Change Voyage" is no longer used and there is no discussion on the web page about the changing nature of the Arctic. Instead the tour operator highlights other aspects of the tour by stating,

> Jewels of the Arctic combines the best of Spitsbergen and Greenland, with a taste of Iceland. Spitsbergen's rugged northwest coast offers mountains, tundra and fjords. While crossing the icy waters of the Greenland Sea, our onboard experts share their knowledge of the region's plants, animals and history. Greenland's remote east coast shows off the immensity of the icecap, fantastic icebergs and a fairytale landscape of granite spires rising 2000 m above exquisite fjords! The local Inuits welcome us, perhaps demonstrating their paddling skills and sharing their amazing culture.

Similarly in 2008 tour operator KMJ11 promoted a tour entitled "Climate Change Challenge Mt Kilimanjaro Mission" whereby the climb of the mountain would involve documenting the effects of climate change. This tour is not being currently offered and the terms "climate change" or "global warming" is not being used anywhere on the tour operator's web page. Instead the tour operator provides information about their various philanthropic activities in developing countries and information for customers on how to minimise their carbon footprint. Similarly, the web page of the tour operator GRE2 demonstrates their commitment to environmental protection and provides advice for customers on how to reduce their carbon footprint, rather than highlighting aspects of global warming. Table 9.5 summarises the minor, moderate and major coverage given to global warming by the operators in both 2008 and 2011.

The mention of the impact of global warming on the tour operator websites may reflect the transition in advertising to design promotions to appeal to the emotions (the non-rational), particularly in electronic advertising (radio, television and Internet) (Beasley and Danesi 2002). Tour operators with minor coverage of global warming appear to be using "scare copy techniques" which are often "designed

*Table 9.5* Coverage of climate change by number of tour operator in 2008 and 2011

| *Level of coverage given to global warming* | *Great Barrier Reef* | | *Greenland* | | *Mount Kilimanjaro* | | *Total* | |
|---|---|---|---|---|---|---|---|---|
| | *2008* | *2011* | *2008* | *2011* | *2008* | *2011* | *2008* | *2011* |
| Minor | 1 | 1 | 2 | 2 | 9 | 8[a] | 12 | 11 |
| Moderate | 6 | 5[a] | 0 | 0 | 2 | 2 | 8 | 7 |
| Major | 2 | 1[b] | 4 | 4 | 1 | 0[a] | 7 | 5 |

a In 2011, one tour operator in each cell no longer mentions global warming
b In 2011 one tour operator no longer operates

to promote services such as insurance as it tries to evoke fear and/or impending disaster" (Beasley and Danesi 2002: 18). The moderate to major coverage tour operators in 2008 introduced discussion of global warming within the context of the environment to create understanding of the natural regions they visit, and, the major coverage tour operators took the opportunity to derive specific tours to meet the new interest in such travel, including the introduction of climate protection initiatives. In 2011 the tour operators reduced mention of these aspects as the market for such tours appear to have diminished for the time being.

## Conclusion

The key finding from this study is that, despite significant media hype surrounding the emergence of last chance tourism as a new and popular niche market, only a small number of tour operators capitalised on aspects of global warming to promote their tours, representing only 8 per cent of all the tour operators in 2008 who operate on the Great Barrier Reef, on Mount Kilimanjaro, and in Greenland. Thus, most of the tour operators who operated on the Great Barrier Reef, on Mount Kilimanjaro, and in Greenland in 2008 did not mention climate change on their web pages (i.e., 94 per cent of operators on the Great Barrier Reef; 90 per cent of operators on Mount Kilimanjaro; and, 80 per cent in Greenland) which, given that they are selling tours to natural assets where climate information would be quite appropriate, is a surprising omission. Consequently, when the tour operators mention climate change on their web pages, they appear to do so deliberately, for educational, promotional and/or competitive reasons.

The analysis illustrated that in 2008, the existence of tour operators who were heavily exploiting global warming and climate change to attract tourists was mostly "media hype". In other words, particular sectors of the media (see, for example, online news services and travel blogs, such as, Leahy 2008; McKie 2007; Shipman 2007) used aspects of global warming and disappearing natural attractions as interesting and exciting angles for their stories but, in reality, few tour operators were, or currently are, using this aspect to promote their tourism products. The study also found that in 2008, among the tour operators who provided major coverage of aspects of climate change, some designed specialist tours to capitalise on the growing interest in climate change, while others developed some climate protection initiatives for tourists. The latter tour operators may be striving for a quality image that will create customer loyalty (Campu and Yague 2008; Miller 2001).

In 2011 the minor and moderate coverage tour operators have not changed their web pages from 2008. However, the major coverage tour operators in 2011 have made changes to their web pages which, in effect, play down the impact of global warming on the natural World Heritage Sites they visit and instead highlights the educational aspects of the tours. Such a change in focus may reflect the tour operator and customers' apathy, complacency and detachment from the impact of climate change, reflecting that the phenomenon is complex and made more difficult to comprehend with the appearance of contradictory information. Perhaps the tour operators want to avoid their tour locations being viewed as last chance destinations as this may lead to criticism that they are exploiting the last remaining natural aspects of that ecosystem. In addition, these tour operators may recognise that to operate viable businesses over the long term necessitates them returning each year to the same destination, otherwise they would need to seek alternative tour locations and offerings. Thus, to operate successfully they need to highlight other aspects of the natural World Heritage Sites they visit to continue encouraging tourists to purchase these tours. Future research involving these major coverage tour operators would confirm the reasons behind the downplaying of aspects of global warming on their web pages.

In the future, potential traveller awareness and interest may again grow regarding the impact of global warming and tour operators may again be tempted to incorporate aspects of global warming more fully into their web page advertising, and tour offerings, particularly because, as Kendle (2008) asks, "who can blame anyone for jumping on the doomsday bandwagon?" In addition, such promotion may reflect the reality that appealing to the human ego remains a terrific sales tool for almost any product (Salkin 2007). Thus, last chance tourism, appears to exist and to be used by a small minority of tour operators operating in natural World Heritage Sites. Given the increasing awareness of climate change among the general population and the associated perception that the variety and uniqueness of the flora, fauna and natural icons are fast disappearing, future research could determine if a greater number of tour operators increasingly use climate change-related shock tactics to attract tourist attention.

## References

*An Inconvenient Truth* (2006). Dir. Davis Guggenheim. Perfs. Al Gore. Film, Paramount Classics.

Baloglu, S. and Mangaloglu, M. (2001) "Tourism destination images of Turkey, Egypt, Greece and Italy as perceived by US-based tour operators and travel agents", *Tourism Management* 22: 1–9.

Beasley, R. and Danesi, M. (2002) *Persuasive Signs: The Semiotics of Advertising*, Berlin: Mouton de Gruyter.

Booth, W. (2007) "See glaciers, while you can", *The Washington Post*, 23 September, 6.

Cai, L., Card, J.A. and Cole, S.T. (2004) "Content delivery performance of world wide websites of US tour operators focusing on destinations in China", *Tourism Management* 25: 219–227.

Campu, S. and Yague, M.J. (2008) "Tourist loyalty to tour operator: Effects of price promotions and tourist effort", *Journal of Travel Research* 46: 318–326.

Cano, V. and Prentice, R. (1998) "Opportunities for endearment to place through electronic "visiting": WWW homepages and the tourism promotion of Scotland", *Journal of Tourism Management* 19(1): 67–73.

Cavlek, N. (2002) "Tour operators and destination safety", *Annals of Tourism Research* 29(2): 478–496.

Cousins, J.A. (2007) "The role of UK-based conservation tourism operators", *Tourism Management* 28: 1020–1030.

Dann, G.M.S. (1988) "Images of Cyprus projected by tour operators", *Problems of Tourism* 3(41): 43–70.

Dawson, J., Stewart, E.J., Lemelin, R.H. and Scott, D. (2010) "The carbon cost of polar bear viewing tourism in Churchill, Canada", *Journal of Sustainable Tourism* 18(3): 319–336.

Echtner, C.M. (2002) "The content of Third World tourism marketing: A 4A approach", *International Journal of Tourism Research* 4(6): 413–434.

Echtner, C.M. and Prasad, P. (2003) "The context of Third World tourism marketing", *Annals of Tourism Research* 30(3): 660–682.

Fennell, D.A. (2008) *Ecotourism*, 3rd edn, Oxford: Taylor and Francis e-library.

Gossling, S. and Hall, M.C. (2006) "An introduction to tourism and global environmental change in tourism and global environmental change", in S. Gossling and C.M. Hall (eds.), *Tourism and Global Environmental Change: Ecological, Economic, Social and Political Interrelationships*, Oxford: Routledge, pp. 1–34.

Intergovernmental Panel on Climate Change (IPCC) (2007) *Climate Change 2007: Impacts, Adaptation and Vulnerability. Contribution of Working Group II to the Fourth Assessment Report of the Intergovernmental Panel on Climate Change*, M.L. Parry, O.F. Canziani, J.P. Palutikof, P.J. van der Linden and C.E. Hanson (eds). Cambridge: Cambridge University Press.

Kendle, A. (2008) "Doomsday tourism: seeing it before you can't, vagabondish", *The Travelzine for Today's Vagabond*, 13 February, no page given.

Krippendorff, K. (2003) *Content Analysis: An Introduction to its Methodology*, 2nd edn, Thousand Oaks, CA: Sage Publications.

Laws, E. (1997) *Managing Packaged Tourism: Relationships, responsibilities and service quality in the inclusive holiday industry*, London: International Thomson Business Press.

Leahy, S. (2008) "Climate change: Tourism at the end of the world". Online: http://ipsnews.net/news.asp?idnews=40837 (accessed 21 July 2008).

Lee, G., Cai, L.A. and O'Leary, J.T. (2006) "WWW.Branding.States.US: An analysis of brand-building elements in the US state tourism websites", *Tourism Management* 27(5): 815–828.

Markwick, M. (2001) "Postcards from Malta: image, consumption, context", *Annals of Tourism Research* 28(2): 417–438.

McCarthy, M. (2007) "A holiday at the end of the Earth: tourists paying to see global warming in action", *The Independent*, 3 May, no page given.

McKie, R. (2007) "10 wonders of the vanishing world", *The Observer*, 28 January, no page given.

McKay, G. (2007) "Climate changes force travelers to consider sites sooner than later", *Pittsburgh Post-Gazette*, 30 December, no page given.

Miller, G. (2001) "Corporate responsibility in the UK tourism industry", *Tourism Management*, 22: 589–598.

National Tour Association (NTA) (2008) "NTA Tour Operator Membership Application". Online: www.ntaOnline:com/doc_lib/2008touroperatorapplication.pdf (accessed 21 July 2008).

Page, S.J. and Connell, J. (2006) *Tourism: A Modern Synthesis*, 2nd edn, London: Thomson.

Rachman, Z. and Richins, H. (1997) "The status of New Zealand tour operator websites", *Journal of Tourism Studies* 8(2): 62–77.

Reimer, G. (1990) "Packaging dreams: Canadian tour operators at work", *Annals of Tourism*

*Research* 17(4): 501–512.

Salkin, A. (2007) "Before it disappears", *The New York Times*, 16 December, no page given.

Shipman, T. (2007) "Nature's 'doom' is tourist boom", *The Telegraph,* 23 December, no page given.

Smight, T. (2008) "10 great endangered places to see while you can", *USA Today*, 18 April, 3.

Stepchenkova, S. and Morrison, A.M. (2006) "The destination image of Russia: From the online induced perspective", *Tourism Management* 27(5): 934–956.

Tsaur, S.H., Lin, Y.C. and Lin, J.H (2006) "Evaluating ecotourism sustainability from the integrated perspective of resource, community and tourism", *Tourism Management* 27: 640–653.

UNESCO (United Nations Educational, Scientific and Cultural Organization Report) (2007) *Case Studies on Climate Change and World Heritage*, Paris: UNESCO World Heritage Centre.

Wan, C. (2002) "The websites of international tourist hotels and tour wholesalers in Taiwan", *Tourism Management* 23(2): 155–160.

Weaver, D. (2001). *Ecotourism*, Milton, Australia: John Wiley & Sons.

Weaver, D.B. (2005) "Comprehensive and minimalist dimensions of ecotourism", *Annals of Tourism Research* 32(2): 439–455.

Weaver, D.B. and Lawton, L. (2006) *Tourism Management*, 3rd edn, Milton, Queensland: John Wiley and Sons.

Wheeler, B. (1994) "Egotourism, sustainable tourism and the environment – a symbiotic, symbolic or shambolic relationship", in A.V. Seaton (ed.), *Tourism: State of the Art,* Chichester: John Wiley and Sons Limited, 647–654.

Wheeler, B. (2003) "Alternative tourism: a deceptive ploy", in Cooper, C. and Cooper, C.P. (eds.) *Classic Reviews in Tourism*, Clevedon, UK, and Buffalo, NY: Channel View Publications, pp. 227–234,

Xiao, H., and Smith, S.L.J. (2006) "Case studies in tourism research: A state-of-the-art analysis", *Tourism Management* 27(5): 738–749.

# 10 Last chance tourism on the Great Barrier Reef

*Alexandra Coghlan*

## Introduction

Within the context of coastal and marine tourism destinations, coral reefs play an important role as an attraction for day visitors, divers, snorkelers, sailors, and for non-swimmers via glass-bottom boats and underwater viewing chambers (Lück 2008). In some areas of the world, reef tourism is the largest national and/or regional income earner and one of the largest employers (Brander *et al.* 2007). For example, in Tropical North Queensland the tourism industry receives 1.9 million reef visitors, is valued at AU$5 billion, and employs 55,000 people (GBRMPA 2007). Many reefs, however, are under increasing pressure from over exploitation either through tourism or more often through over-fishing, coastal development, sewerage discharge, deforestation and farming (Carpenter *et al.* 2008).

Climate change has added a new category of pressure on reefs. Coral reef ecosystems are highly vulnerable to the effects of climate change, most notably rising sea temperatures and ocean acidification (Hoegh-Guldberg 2008). Rising sea temperatures lead to bleaching disturbances, which involves corals expelling their unicellular algal symbionts (*zooxanthellae*), which provide a large proportion of the corals' nutrients through photosynthesis. These bleaching events are likely to cause chronic stress in many reef areas in the coming decades, and if coral communities cannot recover quickly enough, coral ecosystems are likely to become impoverished (Baker *et al.* 2008) and less aesthetically pleasing to tourists who visit them. Evidence of the impact of bleaching on tourism is apparent in a number of studies examining visitor perceptions and preferences. Findings suggest that climate change represents a serious threat to the quality of the experience offered and ultimately to the long-term competitiveness of reef-based destinations (see Kragt *et al.* 2009; Marshall *et al.* 2009; Miles *et al.* 2009).

As scientific reports of reef degradation become even more alarming, the reef tourism industry is likely to become concerned that actual and perceived ecological degradation of coral reefs will affect tourist visitation (Kragt *et al.* 2009). As such, tourism which focuses on vulnerable coral reefs systems, such as the industry based around Australia's iconic Great Barrier Reef (GBR), is a good example of a potential last chance tourism (LCT) destination. The story of the GBR's tourism industry is not a simple one and the financial and political climate, at any given time, is likely to contribute to how the health of the reef is portrayed to potential tourism markets.

This chapter proposes that several multi-scale trends are occurring simultaneously to influence the status of the GBR as a LCT destination. These include; (i) the scientific community and natural resource managers emphasizing the risk of climate change to the GBR in order to highlight the plight of this ecosystem and secure resources for protection, (ii) at a regional scale, the tourism industry has clearly indicated its concern at claims in the media that the health of the reef is deteriorating, (iii) at a international scale, several examples of the GBR being marketed as a last chance destination were found through media reports in prominent newspapers and magazines, (iv) at the destination scale the views of tourists visiting the reef illustrate the complex relationship between perceived reef health, actual reef health, and visitor satisfaction, and finally (v) a number of external events, such as the 2009 'global financial crisis' have played a significant role in the resilience of the GBR's tourism sector over the last few years. In order to build a comprehensive picture of the development of the LCT phenomenon as it relates to Australia's iconic reef attraction, this chapter examines each of these trends, interweaving international and local popular press reports, academic publications, and results of visitor surveys collected through the Marine and Tropical Science Research Facility (MTSRF) between 2006 and 2010 (see Coghlan and Prideaux 2009 for methodological details on MTSRF surveys).

## Media reports by scientists, natural resource managers and conservation groups

A key interest group in the health of the GBR are the managers and conservation groups who draw attention to the threats facing the reef and advocate for greater protection for this ecosystem. The following excerpt from the *Sydney Morning Herald* (23 January 2009) illustrates some of the types of comments that have appeared in the media by scientists, natural resource managers and conservation groups. In this case the comment came from a spokesperson for WWF Australia in response to increasing concern over water quality issues on the GBR.

> 'If we are to give our iconic reef the best chance of withstanding climate change, we must ensure its water quality is as clean as possible', Mr Heath said. 'It is astounding what we are allowing to happen to the reef'.

Overall, the academic literature does not paint a promising future for coral reefs. The health of coral reef ecosystems is now one of the major concerns within the marine research and management community. There is no doubt that as an ecosystem, coral reefs are highly vulnerable to the effects of climate change. One of the main threats facing the GBR is lowered biodiversity through loss of coral and associated changes in reef ecology. In particular, coral reef systems have become threatened by a number of climate change-related factors, most notably rising sea temperatures and ocean acidification (Hoegh-Guldberg 2008; Anthony *et al.*

2008; Thompson and Doleman 2010). Based upon predictions of global warming rates, bleaching events are set to increase in frequency and intensity. One of the most memorable bleaching events occurred in 1998 when 50 to 95 per cent of coral reefs around the world were affected by bleaching. This will become commonplace within 20 years and bleaching events are expected to occur annually in most tropical oceans over the next 30–50 years (Wilkinson 2002). Bleaching disturbances are likely to become a chronic stress in many reef areas and if coral communities cannot recover quickly, coral ecosystems are likely to become impoverished (Baker *et al.* 2008).

In addition to coral bleaching, the GBR has experienced several outbreaks of crown-of-thorns starfish (*Acanthaster planci*). This is a large starfish which feeds on corals by extruding its stomach out onto the coral to digest the living tissue layer. Areas of coral that have been attacked by crown-of-thorns starfish are easily recognizable as having dead white coral in the middle of otherwise healthy coral. Breakouts of the starfish have become more common on the GBR and may be linked to increased development and eutrophication (Brodie *et al.* 2009). Another significant threat to coral reefs is human expansion and development. In particular the increasing levels of freshwater run-off and the high sediment and pollution loads from cleared land are negatively impacting reef systems. The sediment loads and increased nutrients in the water (causing eutrophication of the water) decreases the amount of light reaching corals, which may cause bleaching and speed up the growth rate of competing organisms such as sponges and algae. Faced with these dire predictions, scientists have stepped up their communication strategies to the general public to foster political will aimed at reef protection. Figure 10.1 is taken from the *Climateshifts* website, a blog about science, climate change, politics, coral reefs, and the environment established by

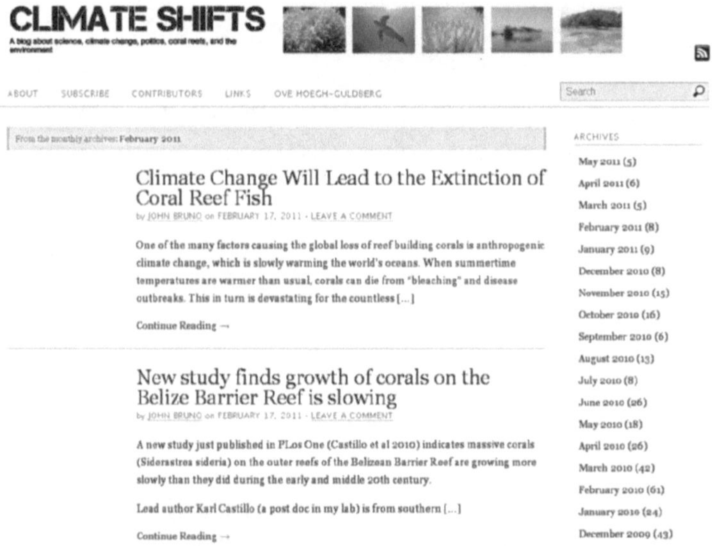


*Figure 10.1* Reports from the blog Climateshifts, illustrating stories of the plight of the reef

Ove Hoegh-Guldberg from the University of Queensland. Headlines from the coral reef category illustrate the information provided by scientists in this highly public arena, which emphases the declining health of the reef and the current and future impacts of climate change.

At the time of writing, another important research institute associated with the GBR, the Australian Institute for Marine Science featured the impact of climate change on reefs and tropical environments more generally on their home page (www.aims.gov.au) (Figure 10.2).

The agency responsible for managing the reef, the Great Barrier Reef Marine Park Authority (GBRMPA) includes climate change impacts as one of its five key focus areas (along with tourism and recreation conservation, heritage and indigenous partnerships, fisheries, water quality and coastal development). *Bleachwatch* is a coral bleaching response plan developed by GBRMPA to monitor and assess the severity and extent of coral bleaching on the GBR each summer. The data generated are then used to continually improve the system's ability to monitor and also to predict future bleaching events. Moreover, two of the three stated goals of *Bleachwatch* are to communicate the impacts of climate change to the general public, by (i) informing 'the community about coral bleaching and to encourage individuals, tourism operators, organisations, and fishers to participate in BleachWatch' and (ii) 'use coral bleaching as a means to communicate the broader impacts of climate change on the Great Barrier Reef' (GBRMPA, date unknown). Results are communicated to the businesses and communities

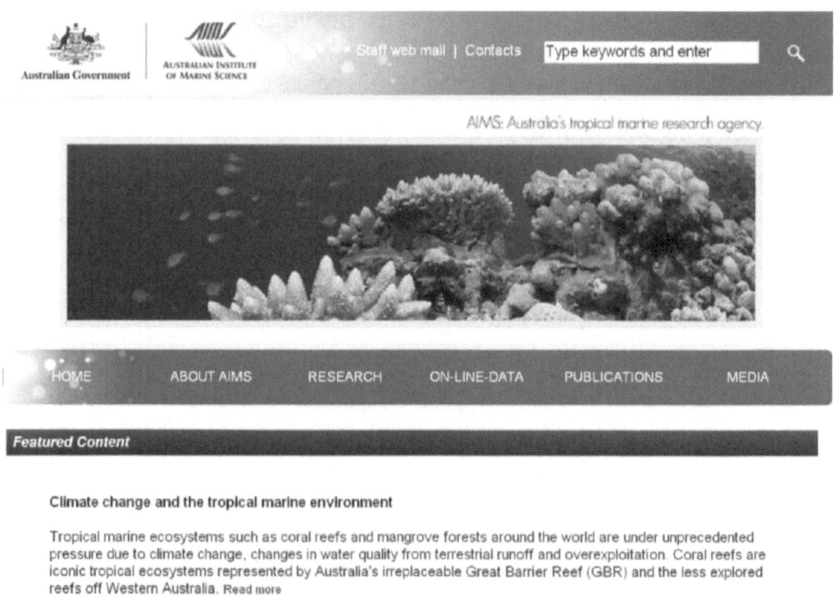


*Figure 10.2* The Australian Institute for Marine Science website homepage featuring the impact of climate change on reefs and tropical environments (image courtesy of the Australian Institute of Marine Science)

that depend on the reef, through coral bleach status reports available on the GBRMPA's website.

The types of communication strategies outlined above and adopted by reef scientists as well as natural resource managers follow the public partnership and outreach models of science communication. In a sense they target the first three steps of the International Association for Public Participation (IAP2) model of public participation adapted from Arnstein's (1969) 'Ladder of Citizen Participation', that is, to inform, consult and involve. At the same time, such reports also provide source material for the popular press. In the case of the GBR, it is perhaps not surprising that popular media articles chart the story of the decline of what is commonly branded as the world's greatest reef. The following section provides examples of how some newspapers, magazines and destination guides have adopted a LCT approach to the GBR. This will be followed by an examination of some of the ways in which the local tourism industry, a sector dependent upon the reputation of the reef as a high-quality attraction, have responded to LCT media.

## Media reports: branding the GBR as a LCT destination

Media reporting on emerging or sustained crises has an immediate impact on tourism destinations (Ritchie 2004) and media reporting that follows a crisis can in fact 'make or break' any given destination. As a result, management organizations are keenly aware of the importance of preparing strong communications strategies aimed at mitigating any negative impacts. Crisis communication and control often involves providing accurate and consistent information to the public and is aimed to enhance the image of the sector in crisis. Often, however, the sector's response may be slow or uncoordinated, allowing other interest groups to fill the information void for their own purposes (Ritchie 2004).

The LCT phenomenon may be considered a response to emerging or sustained crises such as a climate change. This response has prompted some travel operators and tour agencies to recommend vulnerable destinations to consumers before they disappear (Lemelin *et al.* 2010). The expected result is a short-term increase in visitors, with a longer-term decrease in visitor numbers associated with the disappearance of a resource. While environmental threats are not the only cause for stagnant visitor numbers, adverse media reporting of climate change and declining ecological integrity is also anticipated to affect visitor perceptions. For an industry that effectively supports the economy of entire regional areas, such as tourism in Tropical North Queensland, understanding the effects negative media coverage on tourism businesses is extremely important.

To better understand the concerns and possible impacts that media coverage of coral bleaching events and other threats has had on visitor numbers, a number of studies of GBR media coverage were funded by the Cooperative Research Centre for the GBR World Heritage Area (CRC Reef). Most notably a study by Moscardo and Pearce (2002) followed media coverage of the GBR and its main competitors in reef tourism destinations. They confirmed that not all media coverage has been positive. Moscardo and Pearce (2002) illustrate though that the most negative

*Table 10.1* Newspaper coverage of coral bleaching between 1995 and 2010

| Newspaper \ YEAR | 1994* | 1995* | 1996* | 1997* | 1998* | 1999* | 2000* | 2001* | 2002* | 2003* | 2004* | 2005 | 2006 | 2007 | 2008 | 2009 | 2010 |
|---|---|---|---|---|---|---|---|---|---|---|---|---|---|---|---|---|---|
| The Australian | 0 | 0 | 1 | 0 | 5 | 7 | 7 | 5 | 5 | 13 | 1 | 0 | 1 | 2 | 0 | 0 | 5 |
| Daily/Sunday Telegraph (NSW) | 0 | 0 | 1 | 0 | 4 | 7 | 5 | 2 | 4 | 4 | 7 | 0 | 0 | 0 | 0 | 1 | 0 |
| Sunday/Herald Sun (Vic) | 0 | 1 | 0 | 0 | 0 | 1 | 4 | 1 | 4 | 4 | 0 | 0 | 0 | 0 | 0 | 0 | 1 |
| Courier Mail (Qld) | 4 | 0 | 2 | 0 | 8 | 15 | 9 | 6 | 15 | 15 | 2 | 0 | 1 | 0 | 0 | 3 | 0 |
| The Times (UK) | 0 | 0 | 0 | 0 | 2 | 1 | 1 | 2 | 1 | 0 | 0 | 0 | 6 | 7 | 3 | 5 | 1 |
| The Guardian (UK) | 0 | 0 | 0 | 0 | 0 | 4 | 2 | 3 | 3 | 4 | 1 | 0 | 2 | 2 | 2 | 3 | 14 |
| **TOTAL** | **4** | **1** | **4** | **0** | **19** | **35** | **28** | **19** | **32** | **40** | **11** | **0** | **10** | **11** | **5** | **9** | **20** |

Note:
* data taken from Cater 2004

features portrayed of the GBR included a focus on environmental damage and poor weather.

Cater (2004) tracked the media coverage of coral bleaching impacts between 1995 and 2004 in seven regional, national and international newspapers (see Table 10.1). Noticeably, there was an increase in the number of newspaper articles on coral bleaching after the two major events of 1998 and 2002 (Figure 10.3). We might therefore expect that more frequent coral bleaching events may result in further media attention, thereby fuelling the perspective of a dying or disappearing ecosystem, which would then lead to a perception of a declining tourism attraction. This would be a fairly typical example of the issue-attention cycle model proposed by Downs (1972), which explains the fluctuation of media attention toward environmental issues.

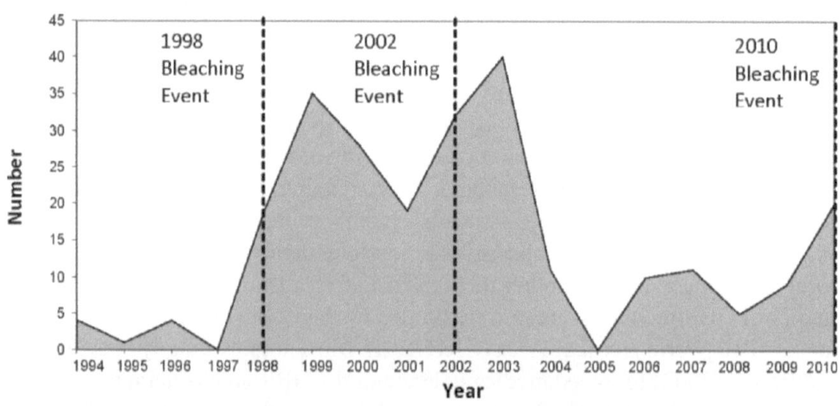


*Figure 10.3* Number of newspaper articles on coral bleaching between 1994 and 2010

This topic has also been covered by travel magazines, which suggest that tourists will soon be faced with an aesthetically less pleasing reef and recommend that tourists visit the GBR 'before it dies'. An example of this is the 'Couldn't find Nemo' article by Salt (2006), which appeared in Cosmos' special travel issue, profiling ten wonders to visit before they disappear. Their article places the GBR among the top ten attractions to visit before it is destroyed by climate change, a clear reference to LCT. The GBR is positioned alongside other destinations and attractions such as Glacier National Park, USA, and Venice in Italy, which are similarly advertised as LCT destinations. Further, *Newsweek* placed the GBR within the top 100 destinations to visit before they disappear (Zakaria 2010). Other magazines such as Qantas' *Australian Way* ran a series of articles on what is being done to protect the reef 'before it is too late' (Southgate 2007).

Based on examples such as these, there is understandably a concern within the industry that 'last chance' media coverage is influencing visitation patterns for the GBR (i.e. increased rates of visitation in the short term and a potential decrease in the long term) as well as raising issues regarding visitor satisfaction and impacts on repeat visitation rates (Baird personal communication 2008). The following section provides an overview of some of the concerns expressed by local industry regarding these reports.

## The local industry's response to LCT on the GBR

Concern over how the deteriorating condition of the reef is portrayed by the scientific community has been voiced by local tourism industry representatives. For instance, Col McKenzie, executive director of the Association of Marine Park Tourism Operators (a trade organization that represents tourism businesses on the GBR), commented in response to media reports of deteriorating reef conditions: 'I do get concerned . . . when I continually hear stories saying the GBR is going to die . . . We have to be really careful that we don't overplay the negative card and not simply from the point of view that it could hurt tourism' (cited in Mercer 2010). Negative reporting of the type illustrated below provides a clear indication of how media reports affect reef tourism and can influence the phenomenon of LCT.

> Tourism operators reliant on the Great Barrier Reef are battling a new menace they say is as damaging to their businesses as crown of thorns starfish. The North Queensland businesses claim publicity about climate change threatening the health of the reef system could have an adverse impact on tourism numbers. Peter Wright, owner of Port Douglas-based Poseidon Cruises and director of the Association of Marine Park Tourism Operators (AMPTO) said that if environmental research continued to dwell on the demise of the Reef, tourism operators might resort to a 'come now or it might be too late campaign'.
>
> (Doomsday talk on Barrier Reef angers tourism operators, *Courier Mail*, 9 September, 2008)

It is also interesting to note that a recent study by Wilson and Turton (2009) on the industry's response to climate change suggests that the local industry has not paid a great deal of attention to *actual* climate change impacts. For instance, the report stated that:

> The majority of operators have only come to believe climate change is happening in the last year and their knowledge of anticipated impact was varied. Some industry practitioners considered climate change to be only another cycle. Of greater concern to the tourism industry is the exchange rate, lack of skills, aviation costs, access, 'rogue' operators and externalities, such as sub-prime collapse in the United States, energy crisis and peak oil, health scares such as SARS, terrorists attacks and war.
>
> (Wilson and Turton 2009: 91)

On the other hand, the report also identified that 'all sectors were concerned with how the region is *portrayed in the media* and how it compares with the rest of the world' (italics added by the author). For example, 'publicity being driven by climate change had convinced the European market that the Great Barrier Reef is already dead' (Wilson and Turton 2009: 92).

Perhaps in part as a response to these concerns, there have been several high-profile tourism campaigns that draw attention to the natural wonders of the GBR. For example marketing campaigns such as the 'Best Job in the World' (below) are reported to have reached an audience of approximately three billion people through the media coverage valued at AUD$400 million. This illustrates a particularly successful campaign designed to draw attention to the GBR and its status as an iconic tourism attraction. In addition, Oprah's 2010 (sponsored) visit to Australia included a trip to the GBR to highlight one of the seven natural wonders of the world; there was little or no mention of how the Reef is at risk of climate change.

> 'The best job in the world campaign has had people from all corners of the globe talking since its launch in January and has become arguably the most sought-after job in the world' [Queensland Premier Anna Bligh] said . . . Tourism Queensland spokesperson Anthony Hayes said earlier this week the competition has already generated more than $110 million in international publicity. 'The answer when things are going tough is you can complain about it or you can be more creative and more aggressive and try to get more people to come' he said.
>
> (Brit wins 'best job in the world', ABC News, 6 May 2009)

## How visitors perceive the reef and the importance of word-of-mouth

Whilst the Association of Marine Park Tourism Operators is clearly concerned about the impact of media reports highlighting reef degradation, Dwyer and Kim (2003) point out that it is not so much the *real* but the *perceived* environmental quality which influence the buying decisions of potential visitors. Indeed, whilst the media portrayal of destination in crisis may have a significant impact on potential visitors, word-of-mouth recommendations by existing visitors can either reinforce media messages or may contradict them; and thus word-of-mouth also has a significant impact on visitation numbers (Yu and Goulden 2006).

In the case of reef tourism, it has been suggested that the relationship between actual reef health and reef tourism demand is not a straightforward one (Anderson, 2007; Kragt *et al.* 2009; Wielgus *et al.* 2002). Existing literature on reef tourism experiences (often diver experiences) has shown that other elements such as clear water, warm temperatures, low health risk, the size of the dive group, and the price of the dive, weather, staff, infrastructure, social group and the variety of experiences on offer are also very important (Coghlan and Prideaux 2007; Fitzsimmons 2007; Rudd 2001; Rudd and Tupper 2002; Shafer and Inglis 2000; Uyarra *et al.* 2005).

As part of a study of divers and trends of GBR tourism (see Coghlan and Prideaux 2007), a series of questions on perceptions of environmental degradation were included in a survey distributed by crew of commercial reef tourism operators to their passengers. The visitor survey focused on socio-demographics, motivations, activities, satisfaction and experiences as well as environmental impact awareness, concern and effects on behaviour. Specifically, these included a question asking respondents to rank threats to the GBR (rising sea temperatures, coral bleaching, crown of thorns, commercial fishing, agricultural runoff, recreational fishing, tourist activity and coastal development), as well as their level of agreement with the following statements on a Likert scale of 1 (strongly disagree) to 5 (strongly agree). (1) The state and health of the Great Barrier Reef influenced my decision to visit it; (2) Reports of future climate change impacts on the reef encouraged me to visit it now; and, (3) I am concerned about my carbon footprint created by visiting the GBR.

A total of 318 usable surveys were collected in 2007, and the results suggest that rising sea temperatures and coral bleaching were not considered the most important threats to the reef (Figure 10.4). Instead commercial fishing was seen

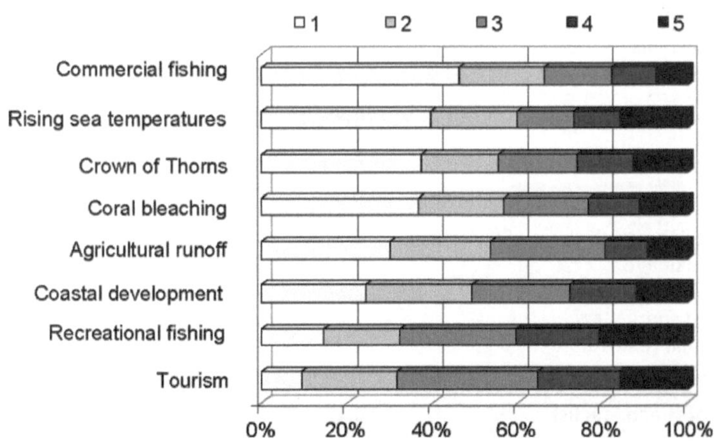


*Figure 10.4* Perceived threats rated by respondents in order of importance from 1 to 5

as the most important threat, perhaps as a result of the recent park re-zoning to increase 'no take areas' in the GBR Marine Park, and the extensive public consultation over fisheries and harvesting impacts at that time. Given these results, it is hardly surprising that less than half of the respondents were concerned about their carbon footprint when visiting the reef, or felt that reports of future climate change impacts on the reef influenced their decision to visit it. However, of interest to the LCT debate, more than half felt that the state and health of the GBR influenced their decision to visit.

Coghlan and Prideaux employed similar questions in another survey conducted after the 2008 International Year of the Reef, which had featured a number of high-profile reports on the state of the reefs around the world. During this period, responses were slightly different from those revealed in 2007. Of the 482 respondents who completed the 2008 survey, 25.5 per cent were able to identify that pollution was impacting reef health, whilst 22.7 per cent mentioned climate change impacts, and only 15 per cent cited commercial activities in the marine park. Domestic respondents were significantly more likely to identify climate

*Table 10.2* Threats to the GBR as identified by respondents

| *People in general* | *All Respondents 15.30%* | *Domestic Respondents 15.18%* | *International Respondents 15.91%* |
|---|---|---|---|
| **Pollution, including** | **25.43%** | **24.82%** | **26.94%** |
| general pollution | 18.07% | 16.31% | 19.69% |
| litter/garbage | 3.02% | 2.98% | 3.40% |
| water pollution | 1.37% | 0.57% | 1.85% |
| oil spillages | 1.51% | 2.41% | 1.11% |
| agricultural pollution | 0.80% | 1.42% | 0.52% |
| land runoff/sedimentation | 0.66% | 1.13% | 0.37% |
| **Climate change, including** | **22.60%** | **25.82%** | **21.98%** |
| climate change generally | 16.37% | 18.72% | 15.77% |
| increasing water temp. | 3.58% | 3.12% | 4.00% |
| bleaching | 1.18% | 1.70% | 0.96% |
| rising sea levels | 0.75% | 1.13% | 0.74% |
| ocean acidification | 0.71% | 1.13% | 0.52% |
| **Commercial activities** | **14.95%** | **11.91%** | **15.25%** |
| tourism | 8.58% | 6.52% | 10.07% |
| fishing | 5.09% | 5.25% | 5.18% |
| mining | 0.05% | 0.14% | 0.00% |
| **Physical damage to the reef structure** | **7.31%** | **7.80%** | **13.99%** |
| physical damage | 3.82% | 2.70% | 4.59% |
| boats/ships/anchoring | 2.88% | 2.27% | 3.26% |
| divers and snorkellers | 2.64% | 1.13% | 3.25% |
| coral/animal removal | 1.23% | 0.43% | 0.07% |
| storm/cyclone damage | 0.85% | 0.99% | 1.41% |
| **Crown of thorn starfish** | **4.10%** | **9.08%** | **1.70%** |
| **Other** | **10.31%** | **5.39%** | **4.23%** |

change/coral bleaching and crown-of-thorns starfish as threats to the GBR, than international respondents ($\chi^2 = 56.457$, $p<0.05$) (Table 10.2).

The change noted between the responses in 2007 (before the International Year of the Reef) and after 2008 suggests that there has been an increased awareness of threats to the reef, perhaps as a result of increased attention to the plight of reefs during that year. There has also been a significant drop in satisfaction scores over the period between January 2007 and June 2010 (when data collection ceased) (Figure 10.5) and a slight change in whether respondents felt that the reef had meet their expectations (Figure 10.6) and whether the trip had met their expectations (Figure 10.7). This may be indicative of an awareness of decreasing environmental quality of the reef, or, as explored in the final section, it may also represent an array of factors that play into visitors' assessment of their experience at the GBR.

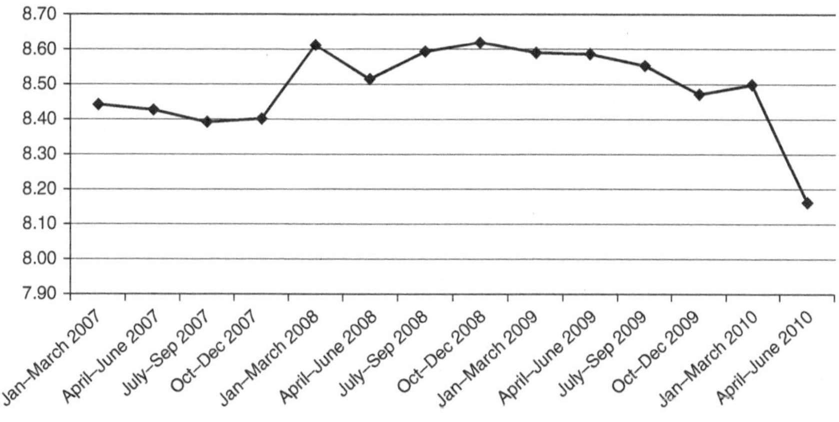


*Figure 10.5* Changes in satisfaction scores (measured from 1 – low, to 10 – high) over four years

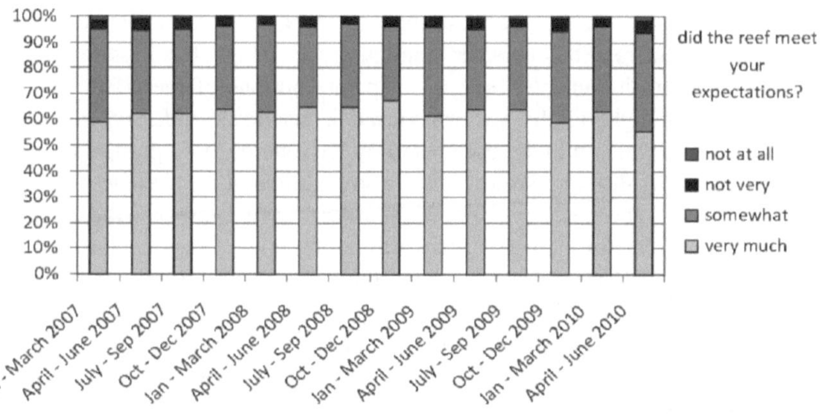


*Figure 10.6* Changes in the extent that respondents felt the reef had met their expectations over four years

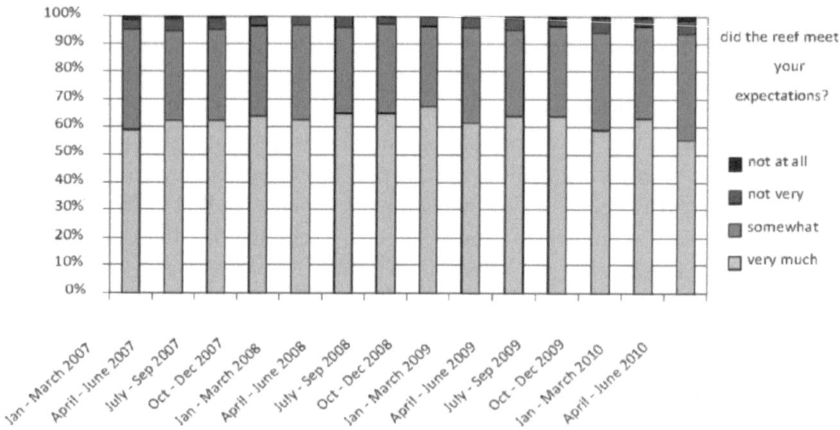


*Figure 10.7* Changes in the extent that respondents felt the trip had met their expectations over four years

Results such as these suggest that visitors to the reef are becoming more aware of potential threats to the reef, whilst at the same time indicating a trend towards decreasing satisfaction with their reef tourism experience. It is not clear however, that the decline in visitor satisfaction is related to increasing reports of reef degradation or the labelling of the GBR as a LCT destination. In fact, returning to the findings by Wilson and Turton (2009), there are a number of other issues that are of concern to the local industry, and these will be explored in the final section of this chapter.

## Tourism on the GBR – additional considerations

Tourism destinations can be described as dynamic places characterised by complex sets of relations between community members, business interests and government. In this chapter, several disparate notions of the GBR have been presented, highlighting the various views of scientists, industry, visitors and popular press, which may be dominated by any one of these groups at different times and across different scales. It has also highlighted the presence and impact of LCT marketing of the GBR. Indeed, in order to fully appreciate the complexity of the GBR's tourism industry, it is worth noting that events unconnected to the health of the GBR also affect the local tourism industry. Some events such as the Global Financial Crisis, an increase in the value of the Australian dollar and cancellation of flights have resulted in falling regional tourism demands.

An earlier review by Cater (2005) of the GBR marine tourism industry suggests that there was a strong correlation between stagnant visitor numbers (mid-1990s to present) and economic factors such as the Asian economic crisis, and social and political instability; for example, September 11. It is worth pointing out that a number of other factors influencing the reef tourism industry and visitor numbers can also be identified. Figure 10.8 highlights that over the last five years a number of natural disasters have occurred both within the GBR region, and more globally, all of which have had consequences for the region's tourist source markets. Four cyclones have been recorded in the region, causing floods, closing roads and airports and resulting in damage to the reef through wave action and marine debris. The examples below illustrate how events such as a cyclone may influence reef tourism by impacting upon the natural resources upon which the tourism depends and by preventing tourists from entering or departing the region.

> The Great Barrier Reef faces an environmental disaster as marine debris is swept out to sea, turning cyclone ravaged Whitsundays into a junkyard. Volunteers are in a race against time to stop tonnes of toxic debris lining the shores being dumped on fragile coral reef ecosystems in the world-renowned Whitsunday islands by a high tide. Experts fear that marine life including fish, turtles and dugong could be choked by the vast armada of post-cyclone rubbish.
>
> (Cylcone debris threat to Great Barrier Reef, *Courier Mail*, 24 March 2010)

> Authorities in North Queensland are keeping a close watch on a low pressure system off the coast after flooding rains caught many people by surprise in the Whitsundays. More than 250 millimeters of rain fell on the region, with more heavy falls expected today. All flights in and out of the Whitsunday Airport have been cancelled due to the flooding, which has cut road access to Proserpine and Airlie Beach.
>
> (Caution urged after flash floods hit N Qld., ABC News, 14 January 2008)

Figure 10.8 draws attention to the number of additional factors that might influence the number of tourists visiting the GBR, adding complexity to the issues surrounding the environmental degradation of the reef. The use of the Internet as a source of information about the destination has continued to expand between 2006 and 2010, possibly indicating a greater use of travel blogs, such as Trip Advisor (Coghlan and Prideaux, in press). This is a trend is being monitored by researchers across the tourism industry for its potential ability to shift marketing channels from formal and content controlled, towards more informal methods including electronic medium and word-of-mouth approaches (Schmallegger and Carson 2008). Whilst this does not yet appear to play a large part in tourism to the GBR, it is certainly a trend worth tracking in the future.

## Conclusion

Using the GBR as an example, this chapter considered a diversity of factors that influence the creation of a LCT attraction. There are multiple tensions between the role of scientists, conservationists, and natural resource managers, who seek to draw attention to the plight of the reef in order to protect an ecosystem that

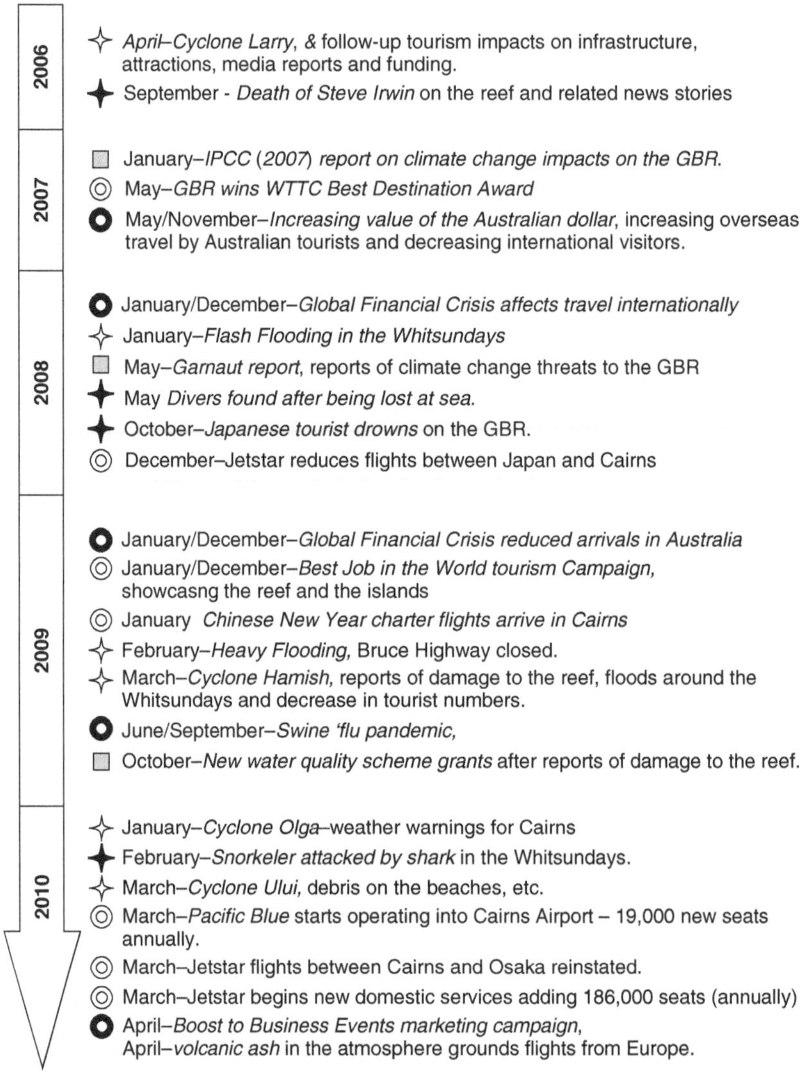


*Figure 10.8* Description of events related to: destination management, natural resource management, localised natural disasters, incidents at the reef, events impacting on sources markets

they (presumably) care about, as well as to position it firmly in the political and funding arenas. The industry, on the other, has expressed concern over the media reports, which they feel portray the reef as 'dying'. The industry also recognises that the environmental degradation of the GBR is but one factor within a number of concerns that might threaten their long-term success. Other factors include changing weather patterns more generally, as well as changes in market trends, transport to the region, and global events and crises that affect travel patterns.

The visitors themselves appear to display some ambivalence with regards to their response to media reports of a dying reef, being largely unaware of the threats and impacts climate change and other factors have on the GBR. Whilst this is not good news for natural resource managers, it has interesting implications concerning the use of LCT as a marketing tool in the region. The research indicates that there is little evidence that the GBR is seen as a LCT destination and the reputation of the reef as the most pristine and greatest reef in the world is still strong as a destination brand. This reveals noteworthy tensions between the media releases by the scientific community looking to foster community support for their work and the popular perceptions of an iconic tourism destination. The relationship between these different dynamics remains to be more fully explored.

## Acknowledgements

This research was conducted with the support of funding from the Australian Government's Marine and Tropical Sciences Research Facility. The author also wishes to thank the operators and their crews for their support in distributing and collecting the visitor surveys.

## References

Anderson, J. (2007) 'The recreational cost of coral bleaching – a stated and revealed preference study of international tourists', *Ecological Economics* 62: 704–715.

Anthony, K., Kline, D.I., Diaz-Pulido, G. Dove, S. and Hoegh-Guldberg, O. (2008) 'Ocean acidification causes bleaching and productivity loss in coral reef builders', *PNAS* 105(45): 17442–17446.

Arnstein, S.R. (1969) 'A ladder of citizen participation', *Journal of American Institute of Planners* 35(4): 216–222.

Baker, A.C., Glynn, P.W. and Riegl, B. (2008) 'Climate change and coral reef bleaching: An ecological assessment of long-term impacts, recovery trends and future outlook', *Estuarine, Coastal and Shelf Science* 80: 435–471.

Brander, L., Van Beukering, P. and Cesar, H. (2007) 'The recreation value of coral reefs: a meta-analysis', *Ecological Economics* 63: 209–218.

Brodie, J., Lewis, S., Bainbridge, Z., Mitchell, A., Waterhouse, J. and Kroon, F. (2009) 'Target setting for pollutant discharge management of rivers in the Great Barrier Reef catchment area', *Marine and Freshwater Research* 60(11): 1141–1149.

Carpenter, K.E, Abrar, M., Aeby, A., Aronson, R.B., Banks, S., Bruckner, A. and Wood, E. (2008) 'One-third of reef-building corals face elevated extinction risk from climate change and local impacts', *Science* 321: 560–563.

Cater, C. (2004) *A Review of the Great Barrier Reef marine tourism industry.* Report prepared for Queensland Tourism Industry Council, Brisbane, Australia. December.

Cater, C. (2005) 'Managing marine ecotourism: the case of the Great Barrier Reef', *International Ecotourism Forum and Symposium*, Zhejiang Forestry University, Lin an City, Zhejiang Province, China, 15–16 November.

Coghlan, A. and Prideaux, B. (2007) *Reef Tourism Quarterly Reports for Townsville, Cairns and Port Douglas, Great Barrier Reef and the Whitsunday Regions*. Quarterly Tourism Barometers produced for the Marine and Tropical Sciences Research Facility, Cairns. Online: www.rrrc.org.au/publications/tourism_barometers.html.

Coghlan, A. and Prideaux, B. (2009) 'Responding to stakeholder research needs using a visitor monitoring survey: the case of the Great Barrier Reef tourism industry', *Tourism in Marine Environments* 5(2/3): 187–199.

Coghlan, A., and Prideaux, B. (in press) *Reef Tourism Drivers and Trends: Synthesis report*. Marine and Tropical Sciences Research Facility, Cairns.

Downs, A. (1972) 'Up and down with ecology – the 'Issue-Attention Cycle', *The Public Interest* 28: 38–50.

Dwyer, L. and Kim, C. (2003) 'Destination competitiveness: determinants and indicators', *Current Issues in Tourism* 6(5): 369–414.

Fitzsimmons, C. (2007) Why Dive? Why Here? A study of recreational diver enjoyment at a Fijian Eco-tourist resort. *Proceedings of the 5th international Coastal and Marine Tourism Congress: Balancing marine tourism, development and sustainability,* 11–15 September, Auckland, NZ.

GBRMPA (date unknown) Online: www.gbrmpa.gov.au/corp_site/key_issues/climate_change/management_responses/bleach_watch2.html (accessed 2 January 2011).

GBRMPA (2007). Online: www.gbrmpa.gov.au/corp_site/info_services/media/media_archive/2007/2007_01_30.html (accessed 3 March 2009).

Hoegh-Guldberg, H. (2008) 'Australian tourism and climate change', *Economic Strategies Pty Ltd.*

Kragt, M.E., Roebeling, P.C. and Ruijs, A. (2009) 'Effects of Great Barrier Reef degradation on recreational reef-trip demand: a contingent behaviour approach', *The Australian Journal of Agricultural and Resource Economics* 53: 213–229.

Lemelin, H., Dawson, J., Stewart, E., Maher, P. and Lück, M. (2010) 'Last-Chance Tourism; the boom, doom and gloom of visiting vanishing destinations', *Current Issues in Tourism* 13(5): 477–493.

Lück, M. (ed.) (2008) *The Encyclopedia of Tourism and Recreation in Marine Environments*. Wallingford, UK: CABI.

Marshall, N.A., Marshall, P.A., Abdulla, A., Rouphael, A. and Ali, A. (2009) *Preparing for Climate Change: Recognising early impacts through perceptions of dive tourists and dive operators in the Egyptian Red Sea.* Gland, Switzerland: IUCN.

Mercer, P. (2010, 21 January) 'Australian scientists and industry tangled up in seaweed', *The National.* Online: www.thenational.ae/apps/pbcs.dll/article?AID=/20100121/FOREIGN/701209811/1002/NEWS (accessed 30 January 2010).

Miles, R.L., Kinnear, S., Marshall, C., O'Dea, G. and Greer, L. (2009) *Assessing the Socio-Economic Implications of Climate Change (Coral Bleaching) in the Great Barrier Reef Catchment.* Rockhampton: CQ University, Institute for Sustainable Regional Development.

Moscardo, G. and Pearce, P. (2002) 'The Great Barrier Reef Media Monitoring Project Report 5: May - August 2001', Supplementary Report on Internet Users. CRC Reef Research Project B2.1.2 Data Summary Report. Online: www.reef.crc.org.au/research/sustainable_tourisim/results/b212.html (accessed 17 November 2008).

Ritchie, B.W. (2004) 'Chaos, crises and disasters: a strategic approach to crisis management in the tourism industry', *Tourism Management* 25: 669–683.

Rudd, M.A. (2001) 'The non-extractive economic value of spiny lobster, *Panulirus argus*, in the Turks and Caicos Islands', *Environmental Conservation* 38: 226–234.

Rudd, M.A. and Tupper, M.H. (2002) 'The impact of Nassau grouper size and abundance on scuba diver site selection and MPA economics', *Coastal Management* 30: 133–151.

Salt, D. (2006) 'Last Chance to See!' *Cosmos* 12: 49–63.

Shafer, C.S. and Inglis, G.J. (2000) 'Influence of social, biophysical and managerial conditions on tourism experiences within the Great Barrier Reef World Heritage Area', *Environmental Management* 26(1): 73–87.

Schmallegger, D. and Carson, D. (2008) 'Blogs in tourism: Changing approaches to information exchange', *Journal of Vacation Marketing* 14(2): 99–110.

Southgate, L. (2007) 'How we keep the reef great', *The Australian Way.* Qantas inflight magazine, January.

Thompson, A. and Dolman, A. (2010) 'Coral bleaching: one disturbance too many for near-shore reefs of the Great Barrier Reef', *Coral Reefs* 29: 637–648.

Uyarra, M., Cote, I., Gill, J., Tinch, R., Viner, D. and Watkinson, A. (2005) 'Island-specific preferences of tourists for environmental features: implications of climate change for tourism-dependent states', *Environmental Conservation* 32(1): 11–19.

Wielgus, J., Chadwick-Furman, N., Dubinsky, Z., Shechter, M. and Zeitouni, N. (2002) 'Dose-response modelling of recreationally important coral-reef attributes: a review and potential application to the economic valuation of damage', *Coral Reefs* 21: 253–259.

Wilkinson, C. (2002) *Status of the coral reefs of the world,* Townsville, Australia: Australian Institute of Marine Science.

Wilson, R. and Turton, S. (2009) 'Cairns case study', in S. Turton, W. Hadwen, and R. Wilson (eds.) *The Impacts of Climate Change on Australian Tourism Destinations: Developing adaptation and response strategies – a scoping study*. Australia: CRC for Sustainable Tourism.

Yu, L. and Goulden, M. (2006) 'A comparative analysis of international tourists' satisfaction in Mongolia', *Tourism Management* 27: 1331–1342.

Zakaria, F. (2010) '100 Places to remember before they disappear', *Newsweek.* Online: www.newsweek.com/feature/2010/100-places-to-remember.html (accessed 12 March 2011).

## Section III

# Ethics, culture, and management

# 11 *Après moi le deluge*

## Ethics, empire, and the biopolitics of last chance tourism

*Mick Smith*

### Preliminaries

Thinking critically about the ethics of tourism involves more than just passing judgements on the actions of individual travellers or producing codes of conduct. Such approaches fail to address more fundamental issues concerning the relationships between ethics *as such* and tourism – understood in its specifically modern (i.e. historically, socially and economically particular) forms. Ethics *as such* can be considered a mode of relating to the world and things in it (a possibility all ethical beings have) such that we concern ourselves with sustaining those things for who/what they are, rather than for their use to us. (In Kantian terms this is to treat them as 'ends in themselves' rather than as 'means' to our ends (Smith and Duffy 2001).) Fundamentally speaking, being ethical is not a matter of following moral norms or rules of conduct at all (however 'modern', 'enlightened' or 'universal' they may claim to be) but of expressing an ability to respond to others in ways that seek to sustain a kind of 'open community' with them.[1] This, we should note, would be true whether that open community is understood solely in human terms or as a wider ecological community.

This last point draws our attention to the ways in which the core values and assumptions of modern Western society have, in fact, often limited ethical consideration to those human beings regarded as being part of our 'political' community, that is, to our fellow citizens. And while those of a more cosmopolitan bent (such as Kant) think that this community should be understood 'globally' as including *all* human beings, the idea that we might relate ethically to things other than humans (such as shrews, trees, or rivers) is often ruled out of bounds. (Kant, for one, explicitly denied such a possibility.)

The dominant presuppositions associated with modernity make it difficult to even imagine an ecological ethics. Ecology, we might say, exposes the ethical limits of modernity: it reveals how this historically particular form of social and economic existence can draw boundaries around and close down any ethical sense of 'community' with the natural (that is to say, more-than-*just*-human) world. In modernity, nature is ultimately regarded as only a means for human ends, a 'resource'. Insofar as tourism is a product of (a Western and now globalised) modernity, then ecology also exposes the ethical limits of even this, explicitly 'cosmopolitan', industry.

What is more, ecology also reveals the destructive consequences of imposing such limits: for treating something as a resource not only *makes use* of it but often *uses it up* altogether. Forests are felled, soils eroded, oceans exploited, mountains mined, and rivers diverted as human needs are given precedence – even when they are, strictly speaking, not *needs* but insatiable *desires* generated within these same modern social and economic systems – for a new car, a computer, an exotic holiday, and so on. Nature is made subject to constant transformations and the more-than-just-human beings and things with which we share this finite world are now disappearing so rapidly that travellers must scramble to see them before they are lost forever. If ecology exposes the ethical limits of modernity/tourism, then this 'last chance' scramble exemplifies the absolute limit modern social and economic systems have placed on nature – the untimely and irrevocable end, the extinction, of many constituents of ecological communities.

'Last chance' tourism is not, on this reading, just a cultural adaptation to a deteriorating ecological situation but a facet of a modern industry profoundly implicated in bringing this situation about. To think about the ethics of tourism critically is to consider the nature of these 'implications'. It is to think about how modern travel has, from its very inception, been caught up in drawing dividing lines between 'open communities' and 'resources' and how this relates to its involvement with the social and economic forms most closely associated with modernity, that is, nation states and global capitalism. Modernisation might even be defined as a process of *assuming* (in both the sense of assuming without questioning and assuming the mantle of) the ultimate right to decide where and when nature becomes a mere resource, where and when it is excised from any sense of ethical community. Another way of putting this is to say modernity's social and economic forms assume 'ecological sovereignty' over the natural world. (Smith 2008, 2011).

Why speak of 'sovereignty'? Because sovereignty is the political concept that has long operated as the master key to open or close the doors of (ethical) community. This remains true even though the form sovereignty takes changes over time. For example, in pre-modern Europe sovereign power was invested in an actual monarch, a sovereign ruler such as Louis XV of France, famous for the remark *Après moi le deluge*: After me, the flood.[2] This is usually interpreted as meaning something like 'I'll do what I want no matter what destructive effects it may have on those who come later: The flood can come after me because *I* am the only one who matters.' Understood in this way it signifies a self-aggrandising claim to be the sole and ultimate power able to decide whether others are, or in this case are not, of any (ethical) concern. That is to say, it is a statement of the monarch's sovereignty, which lies precisely in his assuming the power to decide when and where his 'subjects' could be stripped of their ethical/political standing (and subsequently, perhaps even their lives). Sovereign power constantly holds this 'decisive' possibility over a subject population, and this is as true in modern states (where it takes the form of a 'state of emergency') as in a monarchy (Schmitt 2005). Of course, in late modernity, sovereign power tends to be more diffuse and pervasive, no longer founded in the actual person of the monarch but held suspended in the nexus of nation states and global capital. However, theorists such as

Agamben (1998) argue, this only means its power over populations is, if anything, both more intrusive and extensive. In Hardt and Negri's (2000) influential analysis, this modern global and boundless form of sovereignty is, quite appropriately, referred to as *Empire* (see below).

Speaking ecologically, we must go further still. For sovereignty is not only exercised over human populations, but extends over the whole natural world. As global warming melts the polar ice cap, even the floor of the Arctic Ocean becomes subject to claims of sovereignty over oil and mineral 'reserves' and shipping routes. Ecological sovereignty, then, is the key (biopolitical) term referring to the way in which modern social and economic forms *assume* the power to decide when and where more-than-just-human beings and things can be excised from any ethical/ecological community and reduced to mere objects and resources to fuel the ever growing 'needs' and extravagancies of nation states and capitalism. Such decisions echo Louis XV's lack of concern for the future. Any sense of community with nature is abandoned as the flood of destruction unleashed by global capitalism inundates the entire world. *Après moi le deluge* might be the leitmotif of ecological sovereignty.

Perhaps, though, even in the face of such hopelessness, this phrase might also be interpreted in a rather different way, as expressing a sense regret at the ecological (and cultural) losses that *Empire* portends. Here it might mean something like, 'the coming flood will wash away the world I love leaving nothing behind'. For it is this 'tragic', and potentially 'ethical', sensibility that seems present in the motives behind much 'last chance tourism' and may even inspire a sense of resistance to the flood that ecological sovereignty assumes the god-like right to release upon the Earth.

The contention, then, is that the history of modern travel, from 'first exploratory encounter' to 'last chance to see' is profoundly implicated in the ethical and ecological questions surrounding the imposition of, but also disavowal of, 'ecological sovereignty'. Thinking about tourism in this way may help gather together individual actions, social and economic formations, ethics and ecology in a way that both provides a fundamental analysis of their inter-relations and opens critical possibilities for creating a different more ethical and ecological world. The best way to illustrate this argument might be to trace the operations of ecological sovereignty through a particular history of modern travel, from 'first' encounter to (potentially) 'last' chance to see.

## From 'exploration' to 'last chance to see'

In his history of British travel writing in the twentieth century, Mark Cocker (1992) offers a fascinating summary of the adventures of Lieutenant-Colonel Frederick Marshman Bailey in China, Assam and Eastern Tibet. On one occasion, confined to a cave by heavy snowfalls at the beginning his 1913 expedition to the Tsangpo Gorge, Bailey writes:

> My friendliest visitor was a shrew, which came up on my table as I was working. I grew fond of it as a companion, and yet the more I looked at it the more

> I felt it might be a species which had never been seen or heard of before. As a man I wished it a long and happy life; but as an amateur naturalist I felt that the interests of science came first.
>
> (Bailey in Cocker 1992: 29–30)

Bailey's 'friend' and 'visitor' was accordingly transformed from a living being into a dead specimen which did indeed prove to be new to science. It was named *Soriculus baileyi* in his honour. The intimate nature of their encounter obviously proved no barrier to the shrew sharing the fate of the numerous other creatures Bailey shot, trapped, collected and preserved while surveying previously unmapped areas of the Himalayas.

Those convinced that ethical concerns apply only to our relations with other humans can only celebrate Bailey's natural historical perspicuity. How many of us would recognise a new species of shrew even at such close quarters? Still, some might think there is something ethically troubling about his actions, even, or perhaps especially, because of the way that a being he himself describes as a 'friendly visitor' – but who was in actuality resident within the cave – could find itself so ill-treated. If the shrew offered a kind of animal 'hospitality' and 'companionship' to the actual visitor, the climatically trapped explorer, then Bailey's subsequent actions appear, to say the least, to constitute an extraordinary breach of etiquette between 'guest' and 'host'. Indeed, there may even seem to be a kind of 'betrayal' here, for having once described (and presumably experienced) this relationship in ethical terms (friendship, companionship, fondness) what seems so troubling is precisely that Bailey could then make a one-sided and fatal decision to excise ethical concerns from this relationship in the cause of scientific objectivity.

It is worth thinking about this 'incident of travel' a little more deeply. According to Bailey the 'interests of science' demanded the sacrifice of this singular and unique being, its transformation from companion into a novel *object* of biological curiosity. This 'objectification' obviously constitutes a radical change in the relationship between explorer and animal such that the rather mysterious *ends* and *ways* of this cave-dwelling creature are reduced to a *means* of expanding a taxonomic system that determines, classifies, and fixes inter-species relations according to natural historical and anatomical criteria – criteria that are deemed 'objective' precisely because this taxonomy excludes all 'partiality', including all ethical relations (friendship, companionship, fondness). Ethics can have no role in this system for *ordering* the world according to species, genus, and so on. There is no ecological community here. It is not just that science 'comes first' and ethics takes second place, rather the decision to treat the shrew as no more than an object of certain dimensions with a symbolic value as the 'first of its type' is a decision to strip the ethics out of the relationship.

To be sure, Bailey hints at having regrets, perhaps even a pang of conscience. If pressed he might also have tried to justify his decision in (humanistic) ethical terms; for example, in terms of the benefits consequent upon expanding the scope of scientific knowledge. But then the issue is further complicated because Bailey's actions *and* the pursuit of 'scientific' knowledge may also be rather more 'partial' and less 'objective' than they seem to be. We should not forget that in bringing about the death of his animal acquaintance Bailey gained himself a kind of taxonomic 'immortality' as the shrew's 'discoverer'. To have a species named after oneself was a form of elite social and scientific *distinction* (Bourdieu 1998) that was competitively sought by many contemporary explorers. It gave one 'social capital' that indicated one's stature as an explorer and natural historian, of one's being *the first* to discover species previously unknown to Western science. It was also, of course, a form of naming and classifying nature inextricably associated not only with 'scientific advancement' but the spread of European imperialism and modern capitalism (see, for example, Grove 1995; Lafuente and Valverde 2005; Endersby 2008). Collecting, cataloguing, and classifying species were key steps in exploiting a colonised nature that recognised different species potential as commercial resources, as a kind of 'green gold'.

Bailey's explorations and actions, his 'bioprospecting' (Schiebinger 2004), should be seen in this broader context, for he was not simply a neutral scientific observer any more than he was just a tourist. (Nor for that matter was he, as he claims, strictly speaking an *amateur* naturalist, since it was his professional role to collect such specimens.) A recipient of the Royal Geographic Society's Gold Explorers Medal, Bailey travelled where few if any Europeans had gone before, collecting and collating information on a wide variety of topics, some avidly sought after by the British military. In this sense, and like many Western explorers, Lieutenant-Colonel Bailey's 'scientific' interests were always deeply entangled in the exigencies of (in this case the British) Empire, with trade, and/or conquest.

The practice of collecting natural historical specimens constituted an important aspect of the symbolic and material appropriation of 'foreign' places by (knowing and unknowing) agents of colonial enterprise. As Cocker (1992: 29) writes, Bailey regarded surveying and collecting wildlife specimens as 'two equal halves of an exploratory whole . . . part of a systematic mapping project.' (And while Bailey collected biological specimens his companion, Captain Henry Morshead, surveyed the landscape reducing it to a series of objective measurements for scientific, commercial, and military purposes.) The shrew, we might say, was the victim not only of Bailey's individual decision, but also of a kind of colonial *biopolitics* intimately tied to imperialism and the rise of modern (inter)national capitalism. Its death was a biopolitical consequence of Bailey's role as a traveller at the leading edge of a Western capitalist state's attempt to maintain and extend its empire over other natures and peoples.

The claim is, then, that this apparently insignificant incident in a mountain cave is actually illustrative of a much more general mode of encounter with the 'natural' world. It exemplifies modern Western presuppositions concerning 'ecological sovereignty' over the natural world, where the exercise of this sovereignty is understood precisely in terms of being able to make biopolitical decisions to exclude certain beings from ethical consideration, to excise them from any ethically construed community, thereby re-defining and reducing them to 'objects' and to potential or actual 'resources', that is, mere means for that sovereign power's ends (scientific, imperialistic, and/or commercial). We should also note that claims of ecological sovereignty are often closely connected with claims to priority, with being the first to see and to be in a position to make such decisions. Establishing such priority was, after all, the point of Bailey's expedition to the Tsangpo Gorge.

In this context it is important to emphasise that Bailey and Morshead were *by no means* the first to actually encounter the Tsangpo Gorge. It was, of course, always known to, and inhabited by, native Tibetans, who regarded it entirely differently, as part of a sacred landscape. Bailey was just the first modern European to make his mark there as part of a process of expanding and protecting British trade and claims to sovereignty in Northern India. In fact there is a complex history of 'discovery' regarding the gorge and a Tibetan-speaking native of Sikkim named Kintup who had previously been employed by the British 'Survey of India' to make two undercover trips to it in the late nineteenth century (the second journey taking four years). His report was published in 1889 well before Bailey's expedition. Bailey would, in his turn, be followed by other bioprospectors, such as Frank Kingdom Ward, who in 1924 mounted a plant-collecting expedition funded and accompanied by the Earl of Cawdor. Ward's speciality was rhododendrons and his interests were both scientific and commercial, returning with rare and valuable specimens that he thought would adapt well to the climate and gardens of England and Scotland: to this day the 'grounds of Cawdor castle [itself a major tourist destination in Scotland] bloom with rhododendron species from the Tsangpo Gorges' (Baker 2001: 61)

Like Bailey, Ward and Cawdor were also intent upon being the *first* to find a rumoured waterfall in the upper reaches of the Tsangpo Gorge reported to rival the Niagara and Victoria Falls. This fall actually remained hidden from Western eyes until the 1990s. But again, as one might expect, the explorer's interest in waterfalls were quite different from the native Tibetans, for whom they had deeply spiritual associations. For European explorers the mysterious fall's 'significance lay not in its capacity to induce visions or lead to other realms, but simply its height' (Baker 2001: 59), both in terms of its measurement in feet and, symbolically, in terms of its potentially being 'the first' (highest) amongst all waterfalls in the Himalayas.[3]

Such 'objectifying' and ethically reductive explorations continue apace today although the context has certainly changed. If, for Bailey or Ward 'ecological sovereignty' operated under the auspices of the colonial and commercial interests of the British Empire, then contemporary bioprospecting (like travel in general) might be said to operate under and exemplify that now *global* form of economic and political sovereignty referred to by Hardt and Negri (2000) in their account of late modernity as 'Empire'. Here Empire no longer refers to the colonial reach of any specific nation such as Britain but denotes

> a 'de facto' global sovereignty, a tight web of market and governmental power that leaves no genre of human activity outside its purview. Empire is comprised of both constitutional forms and the 'biopolitical' technologies of order that secure and condition them, taking the form of a decentred network of juridical, governmental, and military organs that respond to local crises in a

> rapid and continuous fashion. Empire is characterised by the interpenetration of different realms of human activity proper to postindustrial society [. . . and] new modes of affective and communicative production.
>
> (Shapiro 2004: 290)[4]

Empire is, Hardt and Negri argue, the mode of sovereignty determining how the contemporary world is biopolitically divided and ruled.

While the nature of Empire has subtly changed, from colonial power to a boundless global capitalism, the role of ecological sovereignty remains much the same. Today, as we have already noted, another aspect of this process has come to the fore where ecologically oriented travel is concerned. For if the explorer's appointed task was to be the *first to see* and 'bring home' some aspect of the natural world, thereby accruing both 'symbolic' and 'material' capital for themselves and their imperial employers, then the over-exploitation of the natural world that this process directly fuelled has altered the situation considerably. The process of objectification that marks the excision of ethical relations to the world's other (more-than-human) inhabitants is now party to the destruction of the natural world on an unprecedented scale. Today it is the scarcity of species and habitats due to their treatment as nothing more than 'resources' for states and corporations that adds urgency to bio-prospecting and to ecologically oriented travel in general. Species and varieties must be seen, collected, photographed, identified and classified or simply visited before they and their natural habitats are despoiled. And so (as other chapters in this volume argue) the purpose of travel today, in late modernity, is more likely to concern being amongst the last, rather than the first, to see some creature, species, habitat or place. For many species, the exploration and objectification that mark the expansion of ecological sovereignty have led only to extinction; only type specimens survive mouldering in museum cases.

Ironically, but by no means unexpectedly, given the Empire's global reach, this situation also applies to Bailey's own destination. For less than a hundred years after he 'discovered' the upper Tsangpo Gorge for Western interests, this impressive rift in the Himalayas with its human residents and its diverse (but still only partially catalogued) biological inhabitants is in immanent danger of being submerged under the hydroelectric schemes of the latest form of Empire to claim sovereignty over the region, China.[5] The proposed new dam on the Tsangpo would, it is claimed, produce 38 MW of electricity, twice that of the socially and ecologically controversial Three Gorges Dam (Mcelroy 2000). Local inhabitants will be forcibly 'relocated', spiritual sites and ecological habitats drowned, and even much of the river's water would be diverted into China.[6] The *deluge*, it seems in this case, can be understood quite literally. And while India objects strongly to this particular scheme, since the waters of the Tsangpo become the Brahmaputra, it too has its own plans for massive hydroelectric plants further downstream on the Tsangpo/Siang and its tributaries. If these developments go ahead, vast stretches of the Tsangpo River and Gorge would be reduced to nothing more than a resource, a power plant to feed China and global capitalism's further growth. As in so many places, the first chance a contemporary traveller may have to see this spectacular place and encounter its human and more-than-human inhabitants may also be the last.

## First *and* last things

So given this situation, who would journey to see Tsangpo Gorge today and why? Are the motives and relations to ecological sovereignty of 'last chance' tourists significantly different from early explorers such as Bailey? After all, there is a sense in which being the first traveller to visit has always been associated with also being the last. Historically speaking, adventurous travellers such as Bailey have arrived on the cusp of powerful economic, military, and socio-political and transformations of the cultural and/or natural world they travelled to see (and this applies to more exotic kinds of tourism in general). Indeed, like Bailey, these adventurers often act as agents, and not just the diarists, of such change. Part of the claim their travelogues make is not just that they are recording things seldom or never recognised before about foreign lands but also that they did this before such cultural and/or natural landscapes were transformed forever by the changes that followed in their wake. Last chance 'tourism' is, at least in this sense, nothing new. Sometimes early travellers' accounts might couch these changes in terms of their being at the forefront of a 'civilising process', or of economic and social 'progress', sometimes, if they are at all critical of the effects of the process we now refer to as globalisation, their accounts take on nostalgic or even apocalyptic tones, for the very things they found so exciting, novel, and valuable about their destinations are those about to be entirely lost or transformed beyond recognition.

Travel writings abound with examples. Wilfred Thesiger remarks in *Arabian Sands*, his account of his travels across the deserts of the Empty Quarter just after the Second World War:

> I went to Southern Arabia only just in time . . . if anyone goes there now looking for the life I led they will not find it, for technicians have been there since, prospecting for oil. Today the desert where I travelled is scarred with the tracks of lorries and littered with discarded junk imported from Europe and America.
>
> (Thesiger in Cocker 1992: 244)

Similarly, Norman Lewis remarks:

> In 1950 with the Far East on the brink of irreversible change, it was clear that, if I ever hoped to go there, not a moment was to be lost. Instinct served me well, for no one in these days, or henceforward, making the journey to Vietnam, to Cambodia, Laos or Burma will ever experience a tenth of the enchantment those countries then possessed. . . . the process of the world's impoverishment in the things that really matter to most of us rushes forward at an ever-increasing tempo . . . The greatest catastrophe has been the burning of the trees. When I went to Brazil in 1979, three million hectares of forest had gone in the previous

> year. One car-manufacturing firm, wishing to diversify into cattle, had started a fire . . . that spread through an area as large as Luxembourg.
>
> (Lewis 1989: vii)

Interestingly, as Cocker (1992: 250) remarks, we can find similar sentiments being expressed by even earlier travellers who visited these same areas (for example, by Harry St. John Philby whose travels in Arabia preceded Thesiger's by some 30 years). They too thought of themselves as arriving 'just in time' to witness vanishing forms of life. This is not to say that Thesiger or Lewis are mistaken, that they had already missed the boat of 'authentic' travel in 'unspoilt' landscapes and cultures. (Nor is it simply that places inevitably change over time and that everyone's point of reference is likely to be their own first encounter with a region.) There is no need to enter debates about 'authenticity' or pristine 'nature' to recognise that if the 'last chance to see' becomes a dominant theme in late twentieth- and twenty-first-century travel, this is precisely, as Lewis notes, because of the exponential increase in the rate and extent of economic, social and environmental change which have made the (often homogenising) effects of the spread of capitalism's Empire more immediately apparent. Travellers such as Thesiger and Lewis are *not* just expressing an over-romanticised vision concerning the Bedouin, the deserts, or the Far East, for the changes, and the losses, such as the burning of the Brazilian forests, are real enough. And if Thesiger and Lewis do sometimes seem overly nostalgic for a past age of travel, then they also recognise that their own footsteps (and writings) often instigated the kind of tourism requiring oil wells and automobile factories.

However, even if we recognise that there is no 'innocent' form of travel where Empire is concerned, there are important differences between earlier traveller's claims to priority and 'latecomers' such as Thesiger and Lewis whose writings are suffused with a sense of regret for cultural and natural landscapes that others will never be able to see. Cocker (1992) goes so far as to claim that Thesiger's 'reverence and sensitivity to the lifestyles' (p. 71) of those he visited means he can be seen as 'one of the first in a line of generous, sympathetic travellers' (p. 71). This, I think, exaggerates Thesiger's role, for not all earlier explorers were culturally or environmentally insensitive and not all latecomers sympathetic or generous, but there is an important point here. For those who travel with the knowledge that they may be one of the last to see a particular place, culture, or species, frequently do so because they believe that destination to be of intrinsic (that is more than just instrumental) value. Their destination and what draws them to it is not just a means to their own (or their nation's) future status or fortune. Nor are they concerned with establishing priority or furthering Empire's sway, but focus instead on recording its destructive consequences. In this sense they express a particular ethics of travel that opposes the reduction of cultures and natures to 'objects', 'resources' or 'commodities'.

To be clear: the claim here is not that early exploratory visitors were all 'self-seeking' or 'acquisitive' (although many of them were) while those who come later are entirely ethical (many are not). The issue here is not just one of passing judgement on the moral *character* of various travellers. Rather, we need to connect the absence of an ethic that fosters an 'open community' with the corrosive sway of Empire. And in this context what matters is, ultimately, the difference between a form of travel that assumes the right to take decisions to *excise* ethics in the service of Empire (for example, Bailey's decision regarding the shrew) or, alternatively, one expressing ethical reservations about the assumptions of sovereignty underlying such decisions. The travels of latecomers, such as Lewis and Thesiger, are ethical insofar as they are motivated by and express the prospective loss to the world of diverse forms of life (natural and cultural) consequent upon the globalisation of (ecological) sovereignty. Their 'last chance' travels pay homage to this diversity and, like last-chance tourism in general, seek to oppose the reduction of an environment (or a culture) to a novelty, curiosity, object or potential resource. Indeed, 'last chance' tourists often carry in their heart a residual hope that they will not, after all, be the 'last to see' somewhere and that some things might yet miraculously survive the flood.

This is obviously a generalisation and it is relatively easy to think of cases where those travelling to distant or inaccessible locations to view disappearing landscapes might be motivated by the symbolic capital or kudos it might give them in particular social circles. (And this despite the fact that, in a competitive society where being first is often all that matters, there is rarely anything like the same cultural cachet in being the 'last to see' something.) Indeed, this 'symbolic capital' might even be translated into actual financial capital if the travellers record or film their experiences for a paying audience. In such circumstances, the fact that the journey is now unrepeatable may help sustain this aura of social distinction surrounding the event precisely because they have done something that no one else now can. But while the immanent demise of a destination certainly gives journeys extraordinary symbolic resonance, the ethical emotions associated with that demise are those of pathos, loss and mourning, of having to let go of something treasured for itself, rather than the acquisitive excitement associated with exploration. In any case, being 'last to see' usually only translates into social and financial capital if the destination is already one that is relatively unexplored, dangerous to get to etc. It is only likely to give the traveller a boost to their ego, social standing or bank balance if she can also, simultaneously claim to be one of the first to see, that is, if the moment of discovery and irretrievable loss coincide. The excitement, the kudos, the adventure, the social capital, and the salability are still largely a matter of being 'first to see or do', or at least in being one of a select few.

These points are illustrated in the advertising material of the (relatively few) trekking companies that offer tours to those areas of the Tsangpo Gorge accessible under Chinese sovereignty. These companies do not tend to emphasise the impending dam project; in fact most of them do not mention it at all (Himalayan Respo Trek 2010; Firante 2010).[7] Rather they emphasise the gorge's 'remoteness', its 'mystique', inaccessibility, natural beauty, and it cultural and natural heritage as a sacred Tibetan site and a 'botanist's paradise' (Nature Trail 2010). These are, of course, all factors in why the loss of areas of the gorge would be so devastating but it seems clear that they believe the gorge's key selling point is the destination's novelty and exoticism.

Again, one has to be careful in drawing ethical implications. That experiential novelty is what makes something saleable does not mean that such treks appeal only to self-interested egoists and/or egotists looking to lift themselves above the crowd. These travellers may very well value and wish to sustain the ecological communities, places or cultures they visit for the richness, diversity and wonder they add to a sadly depleted world *and* for future generations to see. But when it comes to *selling* this experience, there is little or no emphasis on the immensity of the impending cultural and ecological losses. For the only commercial advantage in emphasising the tons of concrete, the tunnels blasted and drilled through mountains, the cables, the roads, reservoirs, the heavy machinery, the debris, and so on would be to tout a 'buy it now while stocks last' approach. In other words, the Tsangpo's value *as a saleable commodity* lies in its uniqueness and scarcity in much the same way that any other resource in a capitalist economy becomes more valuable as demand exceeds supply. The difficulties in finding anywhere that can now be described as 'remote', 'un-spoilt', 'wild', or 'sacred' is precisely what gives it commercial value.

Ironically then, what seems saleable is precisely that a trek to the Tsangpo Gorge seems to offer the visitor an experience of being outside of the orbit of global consumer capitalism, that will show them things whose worth is not reduced to dollars, and perhaps even give the impression of journeying back in time to Thesiger's or Bailey's exploratory times. One might say, then, that such treks have commodified an idealised notion of exploration itself. To even mention the immanent damming of the Tsangpo in an advertisement might then risk shattering the illusion of a land that still lies outside of the sway of Empire. Yet Empire is entirely dependent upon creating new 'economic opportunities', which paradoxically includes commodifying even the desire to escape this same (unethical) system of values. Everything is a potential means to Empire's 'endless' growth.

When those *in the business of exploration* do place emphasis on being the 'last' rather than amongst the 'first' to see the Tsangpo Gorge, it usually appears as a coda for an account of rare adventure, a denouement to a narrative of travel that deploys pathos to round off the action-packed excitement, to draw audiences back into the 'real' world and provide a semblance of emotional balance that might otherwise be missing. Take, for example, the following description prefacing an account of the Western kayakers who decided to paddle down the upper Tsangpo Gorge in 2002.[8]

> The gorge was a Himalayan chasm so shrouded in mystery and danger that a legendary waterfall in its depths, sought by explorers for more than a hundred years, was never even photographed until late in the twentieth century. And it was a place, like Everest, that shimmered with mythic lore and menacing superlatives. By some measures, it is the deepest river gorge in the world – three times deeper, with the river tilted eight times steeper, than the Grand Canyon. Running close to a thousand miles from west to east across Tibet at

> roughly 10,000 feet above sea level, the Yarlung Tsangpo enters the gorge and loses almost all of its altitude in a thundering 150 miles. Along the way, it carves a deep channel between the great peaks of Namcha Barwa (25,446 feet) and Gyala Pelri (23,462), which stand only 13 miles apart. To Buddhists, the gorge is the site of mystical portals to sacred realms; some ancient texts say it will be the last refuge of Buddhism when the rest of the planet falls apart. In an age diminished by the belief that there are no great explorations left undone, the Tsangpo Gorge has remained a fearsome, inviolate anomaly. Nobody had ever successfully paddled the 44-mile stretch of the Upper Gorge from the town of Pe to Clear Creek (beyond which waterfalls make the gorge impassable). It is possible that parts of the gorge had never been seen by a Westerner.
>
> (Heller 2010)

Here we have all the hallmarks of exploratory travel, the journey to a remote, exciting, dangerous and mysterious place unseen by modern Western eyes, the attempt to establish priority – being the first to paddle its length – even the obsession with quantification. The subsequent emphasis in Heller's article (and in the film of the events 'Into the Tsangpo Gorge') is very much on the hardships of the journey, fighting the merciless currents of the river, triumphing over nature and adversity, and so on. Of course Empire's sovereignty is already 'established' over this place but such expeditions can still serve to confirm this in other ways; for example, the expedition into this 'wilderness' was, ironically, sponsored by the Chevrolet Avalanche, a North American petrol guzzling 'sports utility truck'.

Then, right at the end of the article, (and the journey) we are told:

> Curious villagers gathered at our camp and told us that something unutterably sad was soon to unfold in Tsachu and PayŸ and all the other hamlets around the Great Bend. The Chinese-controlled government had told them that in the spring these ancient villages would be depopulated, that the residents would be relocated to new towns and farmland outside the gorge. There were rumours of a national park – or a huge hydroelectric project. No one will live here anymore? we asked. They shook their heads.
>
> We hiked out of the Po Tsangpo to the Lhasa-Chengdu highway. It felt like the end of more than just an expedition.
>
> (Heller 2010)

## Après moi le deluge

We seem to have returned to where we began. An ethics of 'last chance tourism' cannot just be about passing judgements on the moral character of those who kayak or trek places such as the Tsangpo Gorge. Some of these people will just be thrill-seekers, others, even extreme sports enthusiasts, may also be deeply concerned at the fate of the gorge. This is not to say that evaluations of individuals are irrelevant, but that the more fundamental ethical issues relate to 'why' this might be the 'last chance' to see the gorge. In this case, as in most others, this is due to the decisions made by a global form of ecological sovereignty to sever any sense of ethical community with those beings and things with which we share the world at the behest of global capitalism and national interests. This system feeds off and depends upon its claim to have the right to continually 'divide and rule' nature. For ecological sovereignty requires that every thing in the world can be put at the disposal of, and transformed for the profit of, Empire. Such absolutism is hardly less autocratic than the claims made by sovereigns such as Louis XV.

The ethical potential of 'last chance' tourism seems to lie in what its 'take home message' might be. If it communicates something of the impending tragedy in losing places such as Tsangpo, then it can inspire resistance to the reductive and destructive processes operating under the auspices of ecological sovereignty. But for this to happen we need both a critical understanding of the ways in which tourism is implicated in Empire and to sustain and create open ecological communities. The danger is that 'last chance' travel becomes the touristic equivalent of 'just in time' delivery systems that now characterise the movement of goods to their markets a moment before they are required. Here the tourist is no more that the final consumer of goods to which they are delivered 'just in time' before that destination's fate is sealed. But an analysis of ecological sovereignty shows that there is nothing inevitable or fate-like about this process; it is a matter of (Empire's claiming the power of) *decision* rather than *destiny*.

Empire's modern claim to ecological sovereignty is just as hollow as pre-modern claims to the divine right of kings such as Lousis XV to (divide and) rule. Sovereign power is not God-given or a fact of 'real life', it is a claim that can be ethically and politically resisted by those who sustain rather than excise ethics from their relations to the wider world. And there is no sense in waiting until after the flood to bemoan our losses, and building a few floating arks will hardly suffice. We are all visitors in this world and still have a 'last chance' to put ethics before ecological sovereignty. And should we be told that this is a hopeless task, it is worth remembering that floods can have very unexpected consequences. After all, only a few years after Louis XV made his claim '*Après moi le deluge*', the French monarchy was washed away forever by people who recognised that community can only truly be constituted from below not decided upon from on high.

## Notes

1 'Open' here refers to the idea that an *ethical* community is *not* a matter of having to hold something in common with others, for example, to all be members of a particular faith, to be rational beings, to be citizens of the same state, to be communists, to have identical moral outlooks, and so on. Rather it denotes an openness to sharing the world with significant others even when we have nothing whatsoever in common with them (Nancy 2007; Smith 2010).

2 The phrase may actually have been uttered by his extravagant mistress Madame De Pompadour.

3 An American and a Chinese expedition would both claim to have been the first to reach the falls in 1998.

4 Today's exploratory expeditions still collect and classify seeds and other biological materials to make (symbolic and material) profits for nation states and/or global corporations

(Shiva 1997). What matters is being the first to 'patent', to claim private ownership over, the progeny of these creations of nature and/or of long associations with native peoples (Shiva, 1997). There is a counter-argument that indigenous people *may* share in the profits gained from bioprospecting and that it provides benefits for society at large and may even aid a 'market-based' form of conservation (Gairola and Biswas 2008) but given the history of modern capitalism and the examples of bioprospecting documented by authors such as Shiva there are good reasons to be sceptical of such claims.

5 China invaded Tibet in 1950 and has violently repressed dissent within its borders. Some figures suggest more than a million Tibetan's were killed by the Chinese between 1950 and 1983 (Cocker 1992: 241) although the exact figure is disputed.

6 *The Daily Telegraph* (Mcelroy 2000) even reported that the Chinese even planned to use nuclear explosions to blast a tunnel through the Himalayas to the proposed site at Muotuo, although this would contravene the international nuclear test ban treaty.

7 Of course it may also be that the companies feel pressured by the Chinese government's attitudes not to make any mention of the dam or they believe the dam projects will not affect the specific areas of the Tsangpo Gorge they travel to.

8 An earlier solo attempt by a Japanese canoeist ended in his death and an American kayak expedition sponsored by *National Geographic* magazine in 1998 also ended in tragedy when one of the four members was swept away and died.

## References

Agamben, G. (1998) *Homo sacer: Sovereign Power and Bare Life*, Stanford, CA: Stanford University Press.

Baker, I. (2001) 'The Thousand-Petalled Lotus: Tibetan Visions of the Tsangpo Gorge', in K. Cox (ed.) *Tsangpo Gorges*, Woodbridge, Suffolk: Garden Arts Press.

Bourdieu, P. (1998) *Practical Reason: On the Theory of Action.* Cambridge: Polity Press.

Cocker. M. (1992) *Loneliness and Time: British Travel Writing in the Twentieth Century*, London: Secker & Warburg.

Endersby, J. (2008) *Imperial Nature: Joseph Hooker and the Practices of Victorian Science*, Chicago: University of Chicago Press.

Firante, (2010). Online: www.firante.com/trips/Tibet/Trekking%20in%20Tibet%20/Namcha %20Barwa%20and%20Tsangpo%20Gorge%20Trek-137.htm (accessed 1 December 2010).

Gairole, Y. and Biswas, S. (2008) 'Bioprospecting in Garhwal Himalayas, Uttarakhand', *Current Science* 94(9): 1139–1143.

Grove, R. (1995) *Green Imperialism: Colonial Expansion, Tropical Island Edens and the Origins of Environmentalism*, Cambridge: Cambridge University Press.

Hardt, M. and Negri A. (2000) *Empire*, Cambridge, MA: Harvard University Press.

Heller, P (2010) 'Liquid Thunder'. Online: http://outsideonline.com/tsangpo/liquid_thunder_1.html (accessed 13 December 2010).

Himalayan Respo Trek (2010) Online: www.himalayanrespotrek.com/tibet/trekking_in_tibet/yarlung-sangpo.php (accessed 1 December 2010).

Lafuente, A. and Valverde, N. (2005) 'Linnaean Botany and Spanish Imperial Biopolitics', in L. Schiebinger and C. Swan (eds.) *Colonial Botany: Science, Commerce and Politics in the Early Modern World*, Philadelphia: University of Pennsylvania Press.

Lewis, N. (1989) *To Run Across the Sea*, London: Jonathan Cape.

Mcelroy, D. (2000) *The Daily Telegraph*, 22 October.

Nancy, J-L. (2007) *The Creation of the World* or *Globalization*, Albany, NY: SUNY.

Nature Trail (2010) Online: http://allnepal.com/tibet/tibet_trekking_tsangpo_gorgenam.php (accessed 1 December 2010).

Schiebinger, L. (2004) *Plants and Empire: Colonial Bioprospecting in the Atlantic World*, Cambridge, MA: Harvard University Press.

Schmitt, C. (2005) *Political Theology: Four Chapters on the Concept of Sovereignty*, Chicago: University of Chicago Press.

Shapiro, K. (2004) 'The Myth of the Multitude', in P.A. Passavant and J. Dean (eds.) *Empire's New Clothes*, New York: Routledge.

Shiva, V. (1997) *Biopiracy: The Plunder of Nature and Knowledge*, Cambridge, MA: South End Press.

Smith, M. (2008) 'Suspended Animation: Radical Ecology, Sovereign Powers, and Saving the (Natural) World', *Journal for the Study of Radicalism* 2(1): 1–25.

Smith, M. (2010) 'Epharmosis: Jean-Luc Nancy and the Political Ecology of Creation', *Environmental Ethics* 32(4): 85–415.

Smith, M. (2011) *Against Ecological Sovereignty: Ethics, Biopolitics and Saving the Natural World*, Minneapolis: University of Minnesota Press.

Smith, M. and Duffy, R. (2001) *The Ethics of Tourism Development*, London and New York: Routledge.

# 12 Bringing the gaze to the masses, taking the gaze to the people

## The socio-cultural dimensions of last chance tourism

*Raynald Harvey Lemelin and Gary Baikie*

### Introduction

The allure of vanishing landscapes and disappearing wildlife has been recognized and discussed in the tourism literature (Lemelin *et al.* 2010; Dawson *et al.* 2010) and throughout this book. However, apart from Johnston *et al.*'s and Smith's discussions in this volume, very little attention regarding the social, cultural and political dimensions of last chance tourism and its impact on local populations has occurred. This is somewhat surprising when you consider that Indigenous people were featured in various ethnographic exhibits, and that their arts and crafts, photographic images and skeletal remains have been displayed in museums and public displays in North America and Europe throughout the nineteenth and twentieth centuries. When these activities fell out of favour, the need to gaze upon these perceived disappearing cultures was replaced by visitation to parks and protected areas and Indigenous communities (Smith 1977). Fuelling this need to travel to these sites, to witness first-hand these disappearing or vanished cultures, were the various images of Indigenous people used in the promotion of these tourism activities (Dawson 2004). In many instances these images, and later movies, promoted a static depiction of the plain Amerindian who, for example, rode horses, hunted buffalo, and lived in tepees (Notzke 2006). In some cases, visitors to South America were often led to believe that local Indigenous people like the Maya and Inca had disappeared and only their monuments remained (Little 2005; Silverman 2002). This was despite the fact that living Mayans and Incas can be found throughout central and South America, with many of these individuals often associated with the tourism industry (i.e., guides, interpreters, service staff) (Silverman 2002).

Our aim in this chapter is to critique the ethnographic exhibitions and cultural displays of Indigenous people, and challenge the notions of disappearing cultures and vanished people. The goals of this chapter are to: (a) present how issues of colonialism, expropriation and repatriation have been or are being addressed by Indigenous people; and (b) demonstrate how in some cases, the tools of past subjugations (i.e., archaeology) and interpretation can also be used to educate visitors and empower local Inuit populations. We will also demonstrate how last chance tourism has the potential to be utilized as a positive influence in some communities. For example, if people are coming to visit the Arctic because they believe it is disappearing, they can through experiential tourism and interpretation be challenged on these opinions. They can also be shown how the communities are dealing with, and adapting to climate changes. Therefore, the experience does not have to be all about doom and gloom, it can also be about strength and resilience. Having worked together for the past three years on various projects associated to the Torngat Mountains National Park Canada and throughout the Canadian north for over 50 collective years, we feel most comfortable discussing examples pertaining to this particular geographical location. In the multiple case study analysis, we highlight the modern appeal of last chance tourism in the Canadian Arctic, including travel by cruise ship passengers to disappearing Inuit (e.g., Sallikuluk Island in Nunastiavut), European (Beechey Island in Nunavut), and Inuit–European (Hebron, Nunastiavut) sites. Our discussion suggests that addressing the history of colonialism at these sites is contextual and that it must involve Indigenous people. While this discussion and case study is rooted in Canada, we recognize that visitors are travelling to other Indigenous destinations throughout the world for similar purposes, we encourage the local managers and stakeholders involved in these tourism operations to discuss these issues from their own local context and cultural perspectives.

## From museums and fairs to parks and communities – the various dimensions of the indigenous gaze

As Corbey (1993) and Stopp (2009) argue the need to see disappearing Indigenous cultures is nothing new especially when you consider that Indigenous peoples, their rituals, dances, as well as re-enactments of cannibalistic acts and head-hunting have been used to entertain visitors at temporary and permanent international ethnological exhibits throughout Western civilization. Perhaps one of the best-known examples of these ethnological exhibits occurred during the fall of 1880 when a small group of Inuit from Northern Labrador in Canada were enticed to travel to Europe and appear in various European cities as living examples of an ancient and vanishing culture (Stopp 2009). All eight members of the group died from smallpox only a few months into the tour (Stopp 2009). Another famous case is that of Minik Wallace (ca. 1890– 29 October 1918). Minik was one of six Greenland Eskimos ‘given’ to the American Museum of Natural History for scientific experimentation by Robert Peary in 1897 (Harper 2001). Of the group, only Minik and one other individual who was subsequently returned to Greenland survived the ordeal. After the death of his father Qisuk, Minik was adopted by the chief curator, William Wallace, and did not return to Greenland until after 1910 (Harper 2001). Upon his return to the U.S., Minik would discover that his father had not been buried as he had been led to believe but that his skeletal remains were actually on exhibit at the museum (Baikie 1993). According to Baikie (1993), this event transformed Minik for he never trusted any people travelling in the north again. It would take nearly 80 years to convince the American Museum of Natural History to return the four adult Inuit (including Qisuk) remains to Greenland, so they could be properly reburied (Harper 2001). Minik never returned to Greenland, and he now lies buried in the U.S.

In several ways Minik's betrayal reflects many similar experiences by Labrador Inuit in Northern Canada, who often witnessed and protested the removal of artefacts, skeletons and skulls from their traditional territories, throughout the early twenieth century (Baikie 1993). In 2010, the government of Nunatsiavut announced that the skeletal remains of 22 Labrador Inuit from the community of Zoar, removed in 1928 by archaeologists from the Field Museum of Natural History in Chicago, would be returned to Nunatsiavut (Mullen 2010, no page). In some instances (like the Zoar case) repatriation is a long and drawn out process. In other cases, skeletal remains and artefacts have never been returned to the people of Nunatsiavut ('our beautiful land'), nor do we know where they are being stored today (Baikie 1993).

By the end of the nineteenth century, the growing popularity of museums and expositions also created a rising demand for Indigenous art and history and the further disposition of cultural artefacts. Indeed, it was during this time that Inca mummies were removed and regularly displayed in museums throughout the world (Keller and Turek 1998). Artefacts as well as entire Indigenous communities were also removed from such national parks as Badlands, Banff, Death Valley, Glacier, Wood Buffalo, and Yellowstone in both the U.S. and Canada (Binnema and Niemi 2006; Burnham 2000; Sandlos 2008). In the early twenieth century, the Ute Nation was removed from their traditional territories in Colorado to make way for the establishment of the Mesa Verde National Park and UNESCO World Heritage Site, and the protection of the park luminaries – the dead Anazazi (the cliff dwellers) (Burnham 2000). These expropriations (a process also known as Aboriginal erasure) were justified on the grounds that these societies were so acculturated that they no longer practice traditional subsistence patterns, nor could they be trusted to take proper care of cultural artefacts or sacred sites (Burnham 2000). A less discussed but nevertheless important motive for this expropriation was that tourism enterprises could only be developed in these areas once the Indigenous peoples had been removed (Binnema and Niemi 2006; Denzin 2008; Nabokov 2004; Spence 1999). In these controlled environments, Indigenous people and some of their cultural activities (i.e., dances, crafts) could be featured in accordance with park management policies. In many cases, these activities reinforced the stereotypical depiction of the plain Amerindian who had been defeated in battle, and was now relegated to cultural events.

All of these activities whether they be ethnological displays featuring Inuit families, removal of skeletal remains and artefacts, photographic displays of plains Amerindians, or the dispossession of traditional lands through the establishment of national parks, all bear witness to the imperialism and colonialism of nineteenth-century nation-states (Bessire 2003; Rydell 1984). Fortunately by the 1930s, ethnological exhibits were increasingly criticized, and found objectionable on various moral grounds (Corbey 1993). The last expropriation of local people for the establishment of national parks in Canada occurred in the early 1970s (Bernier 2009). It would take another 20 years for the U.S. Native American Graves Protection and Repatriation Act of 1990 to be signed into law. The act is significant because it recognizes that objects removed without tribal approval must be returned when requested. It also acknowledged that Indigenous people are not simply the 'objects of anthropological study, but a people with their own culture and customs that must be accorded the respect that they deserve' (Baikie 1993: 11). Similar programs (i.e., the Returned Treasures Program) aimed at the repatriation of Latin American cultural properties have been instituted by the National Institute of Anthropology and History (Instituto Nacional de Antropologia de Historia) and the National Museum of Anthropology (Museo Nacional de Antropología) in Mexico, in the early 1990s (The National Institute of Anthropology and History, no date).

The decline in ethnographic exhibits in North American and European cities was replaced by the visitor gaze in First Nations and Inuit in Canada and the U.S. (Laxson 1991; Notzke 2006; Smith 1977, 1989), the Maori of New Zealand (Carr 2007), Aboriginal people in Australia (Clark 2009), Indigenous people in South America (Belsky 2000), the headhunters of Borneo (Zeppel 1997), the cannibals of Papua New Guinea (Huang and Lee 2010), and the Hilltribes of Taiwan (Wood 1984).

Interest in the vanishing Canadian Arctic has been fuelled by books (see Grant 2010; Wheeler 2010), articles (Powell 2006), media coverage, and travel reports (Lemelin *et al.* 2010). In many cases, these narratives glorify the exploration of North America by European explorers and highlight the loss of the natural heritage, but often overlook the impacts on local populations. When local populations are discussed in these narratives, they are often depicted as victims of climate change. Two cases studies both located in the Canadian Arctic depict how sites of significant cultural and natural heritage are presented and showcased to visitors, often as sites of vanishing heritage (be it natural or cultural). The first case study highlight's the discovery phase of European exploration while downplaying the Inuit culture (past and present), the second case study emphasizes the importance of Inuit culture and philosophies in the management of national parks as living landscapes. In both cases, the opportunity to engage last chance tourist with communities is provided. This engagement is crucial if visitors are going to be informed that while sites may be transformed through climate change, local people are living and adapting to these changes, and will continue to do so in the future. It is for all intent and purposes trying to create opportunities for local communities to benefit from these changes through tourism, while also continuing to adapt to them.

## Searching for the Northwest Passage in contemporary times

Four designated historical sites (e.g., Devon Island Site at Cape Riley, Franklin Wintering Site, Northumberland House and the *HMS Breadalbane*) are collectively known as the Beechey Island Historic Site. The heritage value of this historic site, located in Lancaster Sound near the Inuit hamlet of Qausuittuq 'the place with no dawn' (also known as Resolute, Nunavut), is derived from the archeological evidence of pre-contact Inuit occupation, and its association with the search for the Northwest Passage (Beechey Island is also known as the entrance way to the Northwest Passage), the loss of the Franklin Expedition and subsequent investigations (39 different rescue missions in all) into the disappearance of Sir John Franklin and his men. Archaeological resources include evidence of Inuit occupation, the wreck of the *HMS Breadalbane* (1853), cairns, caches, a cenotaph, six memorials dedicated to Franklin and various other polar explorers, and the gravesites of four individuals, including three crew members from the Franklin Expedition (Powell 2006) (see Figure 12.1). Beechey Island was declared to be a 'site of territorial historical significance' by the government of the Northwest Territories in 1979 (Beattie and Geiger 2000; Canada's Historic Place no date a, no date b). This historic site, according to Stewart *et al.* (2010) has become one of the most popular destinations for cruise ships in the Canadian Arctic. How this increasing interest in Beechey Island is affecting tourism opportunities in the closest Inuit community of Qausuittuq is discussed next.

The Inuit hamlet of Qausuittuq (pop. 229 in 2006), also named Resolute after the *HMS Resolute*, one of the vessels that came in search of the Franklin Expedition (Hamlet of Resolute Bay no date), is situated on the south coast of Cornwallis Island in Perry Channel. The hamlet is home to a number of scientific research projects including the Polar Continental Shelf project, various adventure expeditions to the North Pole, Quttinirpaaq (Ellesmere Island) National Park, and Beechey Island, and a proposed new military training centre (Hamlet of Resolute Bay, no date). Also up until recently, Resolute acted as an entry/exit point in and out of the high north for cruise visitors arriving or departing by jet plane. While no official visitor figures are available, considering that only two individuals visited Quttinirpaaq in 2009–2010 (Brackley *et al.* 2011), the impacts from tourism in the hamlet are most likely quite modest.

Another site, this one located on King William Island (near the Queen Maud Gulf Migratory Bird Sanctuary in the Kitikmeot region of Nunavut) and also

*Figure 12.1* Beechey Island, Nunavut (Photo credit: Parks Canada)

associated to Franklin's last expedition, was found in 1992. Numerous artefacts and potential wood fragments from a lifeboat or sledge along with 'nearly 400 human bones and bone fragments, representing a minimum of 11 men, were also found at the site' (Keenleyside *et al.* 1997: 36). This was not, according to Keenleyside *et al.* (1997), the first time that skeletal remains associated to Franklin's last expedition had been found on King William Island. While the remains of the rest of the crew and the two ships – the *Terror and Erebus* – is still unknown, the ships have been declared as national historic sites and are commemorated on O'Reilly Island (an island close to King William Island) in the Queen Maud Gulf (Rondeau 2010). Concerns over the displacement of artefacts and skeletal remains have been expressed by nearby Inuit communities. It is a common belief that the remains should be left untouched (Communication with community members, 2010 and 2011). Because of these beliefs, Inuit and First Nation citizens are becoming much more hesitant in divulging where these sites might be located to non-locals.

The most recent archaeological find associated to the Northwest Passage, only made possible because of declining ice in the region, occurred in the summer of 2010 when a team of researchers from Parks Canada discovered the wreck of the *HMS Investigator* (abandoned in 1853) and three graves reported to be from the crew near Banks Island, at the western entrance of the Northwest Passage. The '*Investigator* had sailed from England in 1850 under Capt. Robert McClure to join the frantic search for the ill-fated Franklin expedition, entering the Arctic from the western side in hopes of finding Franklin's two ships emerging from the fabled passage' (Martin 2010, no page). According to Parks Canada officials, this is 'one of the most important shipwrecks in Canadian history because the *Investigator* carried Capt. Robert McClure who is the first European to discover the western entrance to the Northwest Passage' (Martin 2010, no page). Considering the popularity of the Northwest Passage and the various rescue missions associated to the ill-fated Franklin expedition, one would expect that this site may become a national historic site and be included in future cruise-ship tourism marketing strategies.

The fact that so much money is being invested in finding these sites of European exploration and that very little discussion of Inuit presence (past and present) in these protected areas is provided, reinforces the colonialistic and imperialistic tendencies (whether this intended or not), and relegates the Inuit people to historical footnote or archaeological curiosity. Ironically, the opportunity to find these sites is being provided by climate change, thereby illustrating like Johnston *et al.* (this volume) suggested, how last chance tourism can also become opportunities for the first chance to see. As we discuss next, the exclusion of the Inuit people in the Beechey Island complex does not need to be so.

## Torngat Mountains National Park and the Hebron national historic site Nunatsiavut Newfoundland/Labrador

Founded in 2005, Torngat Mountains National Park (TMNP) extends from Saglek fjord in the south and north to Killniq Island, from the provincial boundary of Quebec in the west to Labrador Sea in the east. TMNP has the highest mountains in Canada east of the Rocky Mountains, glaciers, fiords, and a variety of wildlife including caribou, seals, whales, raptors, and waterfowl (National Geographic 2011). Polar bears and a large subspecies of black bear (Barren-land variety) are found throughout the park (Veitch and Harrington 1996). The gateway communities for the park are the two Inuit settlements of Nain Nunatsiavut and Kangiqsualujjuaq Nunavik. While staff for the park are located in both communities, Nain is the central administrative centre for TMNP.

TMNP is a testament to an ancient and living Inuit culture containing hundreds of archaeological sites including tent circles, sod houses, food cashes, burial sites, and *Aullâsimauet* (Inuit settlement camps). More recent developments also being recognized as historically significant include the Moravian missions at Rama and Hebron, Hudson Bay posts at Saglek Bay et Nachvak fjord, and military installations (a German meteorological station on the Hutton peninsula from World War II, and two radar posts from the cold war). Despite being one of Canada's newest national parks and the newest Arctic transboundary area, TMNP in its short history has implemented a cooperative management board composed entirely of Inuit, developed a management approach to national parks that is cognizant of Inuit knowledge and philosophies, and increased its visitation to 461 in the 2009–2010 season (Brackley *et al.* 2011).

Due to its proximity to kANGIDLUASUk (the base camp for the park), spectacular scenery, two sod house villages, over 683 gravesites and above-ground stone burial chambers, and a mass reburial site of 113 repatriated skeletal remains, Sallikuluk Island (Rose Island) (Figure 12.2) is one of the most popular destinations in TMNP (Baikie unpublished data). The site is particularly important because it was used as a whale harvesting site. Here whales would be trapped at low tide, and processed for food, oil, and shelter.

Sallikuluk is attracting an increasing number of visitors from the base camp and cruise ship expeditions (Parks Canada 2010). Protecting the gravesites and burial chambers from disturbances caused by humans and polar bears are two of the most pressing concerns with the management of this area. While little can be done to control polar bear disturbances, visitors are asked to refrain from disturbing any site that resembles a historical burial ground and leave all artefacts on site. Furthermore, all visits are currently supervised by Inuit guides. Ongoing concerns over these activities compelled the TMNP co-operative management board to develop a management strategy for Sallikuluk Island in the park management plan, this includes completing the site inventory, establishing a monitoring strategy, documenting the oral history of the site, developing guidelines and protocols for site visitation, and training local guides and interpreters (Parks Canada 2010).

Located outside of TMNP and down the coastline is the abandoned community of Hebron. Founded in 1831 by Moravian missionaries, Hebron was inhabited until 1959, when 58 Inuit family were removed and relocated throughout Inuit settlements in southern Labrador. Many of these resettled families remained an underprivileged minority of these communities (e.g., Nain, Hopedale) (House 1989). Hebron is a historic site partially managed by Parks Canada and the

*Figure 12.2* Sallikuluk Island, Nunatsiavut (Photo credit: Harvey Lemelin)

government of Nunatsiavut (Maher and Lemelin 2011). The Maritime Archaic, Pre-Dorset, Dorset, Labrador Inuit and Moravian presences here are highly visible in the form of sod houses and the mission building. Several large sod house foundations and above-ground stone burial chambers attest to a pre-Moravian and pre-Inuit occupation of the site (Loring and Arendt 2010).

The Moravian mission at Hebron was once composed of the main mission building (built between 1833 and 1837) which housed the chapel as well as the living and working apartments for the missionaries (Loring and Arendt 2010). Also found nearby was the Hudson's Bay Company store (and post office), a boat house and seal-oil plant, garden shed, goat barn, powder house, residences for teachers, RCMP and Hudson Bay personnel, and the 'God's Acre' cemetery where several Inuit gravesites (pre-Christian and post-Christian) and European gravesites are located (Loring and Arendt 2010). The site also consisted of an impressive network of gardens, raised gravel paths, aqueduct, and bridges (Loring and Arendt 2010). The 68 Inuit families located at Hebron during the Moravian era lived in various dwellings including skin tents, umiaks, sod-covered wooden-frame cabin 'combining traditional house construction elements (sod-walls for insulation, south facing window, and porch configuration) with introduced European influences (wooden walls and wood-burning stoves), and wooden-houses' (Loring and Arendt 2010: 37). Today only the main mission building, the cemetery, some ruined mid-twentieth-century houses used by the RCMP and by the Hudson Bay personnel and a portion of the seal-oil rendering plant are still standing (Figure 12.3).

Efforts to protect the site resulted in the Taipsumani Archaeological Project in 1990. This joint project between the Torngasoak Cultural Centre and the University of South Carolina's archaeology department was held in Hebron in part because there was some interest in renovating the site and potentially developing an eco-lodge there (Loring and Arendt 2010). The fieldwork portion of the archaeological assessment, conducted under the auspices of Gary Baikie formerly at the Torngasoak Cultural Centre in Nain and now a Parks Canada employee, was aimed at encouraging Inuit youth to participate in archaeology research (Loring and Arendt 2010). Almost ten years later, on the fortieth anniversary of the Hebron relocation, 'Torngasok, the Nunatsiavut Government, and the Labrador Inuit Health Commission arranged for a reunion of the Hebron families and their descendants at Hebron . . . The reunion was a powerful reaffirmation of Inuit community integrity and the continuity of social and emotional bonds to their north coast of Labrador homeland.' (Loring and Arendt 2010: 54). Following the creation of TMNP in 2005, the premiere of Newfoundland and Labrador, Danny Williams, offered his excuses to the victims of the Hebron relocation and unveiled a plaque commemorating these individuals.

Hebron's historic prominence has made it an attractive destination for Inuit subsistence hunting and fishing parties, history buffs, visiting researchers, as well

*Figure 12.3* Hebron, Nunatsiavut (Photo credit: Harvey Lemelin)

as tourists (Loring and Arendt 2010; Maher and Lemelin 2011). Concerns regarding visitor conduct and the quality of the experience provided at the site resulted in the 2009 creation of a monitoring programme resembling the "Watchmen" programme instituted at Gwaii Haanas National Park in British Columbia. The TMNP programme consisted of an Inuit family (two parents and two children). The goal was to have this family oversee the protection of sensitive heritage sites, educate/interact with visitors, and promote the living culture of the Inuit people. A similar programme is now being suggested at Ramah (another Inuit and Moravian settlement located further north in the park) (Maher and Lemelin 2011).

The Sallikuluk Island management strategy and the Taipsumani Archaeological Project demonstrate that archaeology, a science often vilified in Northern Labrador, can, with the co-operation of local people and organizations, provide opportunities for community participation and involvement, and affirm 'an Inuit voice and perspective in deconstructing the narrative of the historical period which, too date, has primarily been shaped by the voluminous records and accounts of the Moravian missionaries' (Loring and Arendt 2010: 33). In the case of Nunatsiavut, these projects provide an opportunity for the Inuit to empower themselves and tell their own version of these events, while also showcasing their living culture to visitors.

Unlike the previous case study in Nunavut, where the people of Resolute are largely peripheral to existing management and interpretation strategies in Beechey Island, the Inuit of Nunatsiavut and Nunavik are central to tourism and management strategies in the national park and historic sites. By becoming agents of interpretation and management, they have been able to address the issues of colonialism and expropriation in these tourism destinations. They have also, as Johnston *et al.* noted earlier in Chapter 2, transformed the last chancc to a first chance opportunity.

## Conclusion

By providing a historical overview of ethnological displays and Inuit people, the public display of skeletal remains, and the expropriation of artefacts and lands of Indigenous peoples, we demonstrated how these issues affected Indigenous peoples and illustrate how Indigenous people are addressing these issues through repatriation and, in the case of TMNP, innovative parks management approaches. In the case of Sallikuluk Island and Hebron, archaeology and interpretation is now being used to educate visitors and empower Inuit. Although the cultural dimension of LCT may not be a prime motivator for many visitors, TMNP staff and management have capitalized on this attraction through various educational and interpretative measures, emphasizing that the Inuit culture is not vanishing, it has adapted, and it will continue to do so.

In the case of the Beechey Island historic site, much of the current management strategies have been aimed at protecting and promoting the heritage of Europeans and Euro-Canadians. Indeed, virtually no discussion is provided of the Inuit legacy prior to the European 'discovery' of this site. Visitors to this historic site are in essence presented with the death of European explorers and the virtual erasure of the Inuit people from the area. In this context, the opportunities provided through last chance tourism are not being capitalized upon. These challenges aside, the decline of the in-situ gaze of Indigenous peoples in ethnographic displays has not, according to some Indigenous leaders, completely disappeared. For countless artefacts, cultural and skeletal remains continue to be the property of museums and private collectors (Baikie 1993). Repatriation efforts as the case of Minik, Sallikuluk Island and most recently the Zoar case in Nunatsiavut are long and difficult processes. Despite a legacy of colonial hegemony, public expositions and expropriation, silencing and dismissal, Indigenous peoples continue to display goodwill and a willingness to work and educate visitors who may have been attracted to the area by concerns of disappearing natural heritage, or in some cases vanishing Indigenous people. By recognizing and discussing past wrongdoings, Bowman and Pezzulo (2010) argued that these sites can be transformative and restorative, that these sites can help assess damage, educate key constituents, empower locals, and galvanize much needed support. It is therefore not just about preserving history, but also about shaping the present and the future. By doing so, we will be able to heal the hurt, and move forward and learn from various tourism opportunities, including last chance tourism.

## Acknowledgements

We would like to thank the Social Sciences and Humanities Research Council of Canada and Parks Canada for their funding support. We are thankful for the photographic contributions of Garry Enns, External Relations Manager, Nunavut Field Unit – Parks Canada.

## References

Baikie, G. (1993) 'What do the Labrador Inuit want?' *Inuit Art Quarterly* 8(3): 8–16.

Beattie, O., and Geiger, J. (2000) *Frozen In Time: The Fate of The Franklin Expedition.* New York: Sterling.

Belsky, J. (2000) 'The meaning of the Manatee: An examination of community-based tourism discourse and practice in Galez Point, Belize', in C. Zerner (ed.), *People, Plants, and Justice: the politics of nature conservation.* New York: Columbia University Press.

Bernier, L. (2009) *La bataille de Forillon.* Chicago: Fides.

Bessire, L. (2003) 'Talking back to primitivism: Divided audiences, collective desires', *American Anthropologist,* 105(4): 832–838.

Binnema, T., and Niemi, M. (2006) 'Let the line be drawn now': Wilderness, conservation, and the exclusion of Aboriginal people from Banff National Park in Canada', *Environmental History* 11(4): 724–750.

Bowman, M.S., and Pezzulo, P.C. (2010) 'What's so "dark" about "dark tourism"? Death, tours, and performance', *Tourist Studies* 9(3): 187–202.

Brackley, C., Fick, S., and Greig, K. (2011) 'À la carte: Alternative dimensions: resizing Canada's National Parks by the numbers of visitors', *Canadian Geographic* April: 34–35.

Burnham, P. (2000) *Indian Country, God's Country: Native Americans and the National Parks.* Washington, DC: Island Press.

Canada's Historic Places (no date, a) 'Beechey Island'. Online: www.historicplaces.ca/en/rep-reg/place-lieu.aspx?id=7105&pid=188&h=Beechey,Island (accessed 1 March 2011).

Canada's Historic Places (no date b) 'H.M.S. Breadalbane'. Online: www.historicplaces.ca/en/rep-reg/place-lieu.aspx?id=15149 (accessed 1 March 2011).

Carr, A. (2007) 'Maori nature tourism businesses: Connecting with the land', in R. Butler and T. Hinch (eds.), *Tourism and Indigenous Peoples: Issues and Implications*, 2nd edn. Oxford: Butterworth-Heinemann.

Clark, P. (2009) 'Ripaŋu Yidaki : Aboriginal social frameworks in a Musical Ecotourism Business', *Journal of Ecotourism* 8(2): 176–192.

Corbey, R. (1993) 'Ethnographic showcases, 1870–1930', *Cultural Anthropology* 8(3): 338–369.

Dawson, M. (2004) *Selling British Columbia: Tourism and Consumer Culture, 1890–1970*. Vancouver: UBC Press.

Dawson, J., Stewart, E.J., Lemelin, R.H., and Scott, D. (2010) 'The carbon cost of polar bear viewing tourism in Churchill, Canada', *Journal of Sustainable Tourism* 18(3): 319–336.

Denzin, N. (2008) *Searching for Yellowstone: Race, Gender, Family, and Memory in the Postmodern West.* Walnut Creek, CA: Left Coast Press.

Grant, S.D. (2010) *Polar Imperative: A History of Arctic Sovereignty in North America.* Vancouver: Douglas & McIntyre.

Hamlet of Resolute Bay (no date) 'Welcome to the Resolute Bay, Nunavut Photo Albums'. Online: www.arcticcircle.ca/Baffin/Res/ (accessed 2 April 2011).

Harper, K. (2001) *Give Me My Father's Body: The Life of Minik, the New York Eskimo.* New York: Washington Square Press.

House, D.J. (1989) 'The sustainable outport: A model for community development?' *Canadian Journal of Community Mental Health* (Revue canadienne de santé mentale communautaire), 8(2): 25–40.

Huang, W.J., and Lee, B.C. (2010). 'The tourist gaze in travel documentaries: The case of cannibal tours', *Journal of Quality Assurance in Hospitality & Tourism* 11(4): 239–259.

Keenleyside, A., Bertulli, M., and Fricke, H.C. (1997) 'The Final days of the Franklin Expedition: New skeletal evidence', *Arctic* 50(1): 35–46.

Keller, R.H., and Turek, M.F. (1998) *American Indians and National Parks*. Tucson, AZ: The University of Arizona Press.

Laxson, J.D (1991) 'How "we" see "them": Tourism and Native Americans', *Annals of Tourism Research* 18(3): 365–391.

Lemelin, R.H. (in-press) 'L'aire protégée transfrontalière Torngat-Koroc: Gouvernance et retombé économique sur la péninsule du Labrador', in D. Chartier et T. Martin (eds.) *Parcs Nunavik.* Les Presses de l'Université du Québec. coll. Imaginaire Nord Droit au pôle.

Lemelin, R.H., Dawson, J., Stewart, E.J., Maher, P., and Luck, M. (2010) 'Last chance tourism: boom, doom, and gloom', *Current Issues in Tourism* 13(5): 477–493.

Little, W.E. (2005) 'Mayas in the marketplace: Tourism, globalization, and cultural identity', *Journal of Latin American Anthropology* 10(2): 442–444.

Loring, S., and Arendt, B. (2010) '". . . They Gave Hebron, The City of Refuge . . ." (Joshua 21:13): An archaeological reconnaissance at Hebron, Labrador', *Journal of the North Atlantic* 2(1): 33–56.

Maher, P., and Lemelin, R.H. (2011) 'Northern exposure: Opportunities and challenges for tourism development in Torngat Mountains National Park, Labrador, Canada', *Polar Record* 47(1): 40–45.

Martin, D. (2010) 'Exclusive: Historic Northwest Passage wreckage discovered beneath Beaufort Sea. Parks Canada team finds remains of HMS Investigator', *Calgary Herald.* Online: www.calgaryherald.com/technology/Exclusive+Historic+Northwest+Passage+wreckage+discovered+beneath+Beaufort/3330222/story.html#ixzz1Jcs9ZBlv (accessed 1 March 2011).

Mullen, W. (2010) 'Field museum to return Inuit remains: Skeletal Remains of 22 people were dug up from Canadian cemetery in 1928', *Chicago Tribune*. Online: www.allvoices.com/s/event-6348751/HR0cDovL2ZlZWRzLmNoaWNhZ290cmlidW5lLmNvbS9+ci9jaGljYWdvdHJpYnVuZS9uZXdzL2xvY2FsL34zL0RLYkFhQ1dnaEhZL2N0LW1ldC1pbnVpdC1yZW1haW5zLXJlcGF0cmlhdGVkLTIwMTAwNzE5LDAsNTkxNTY0MS5zdG9yeQ== (accessed 1 March 2011).

Nabokov, P. (2004) *Restoring a Presence: American Indians and Yellowstone National Park.* Norman, OK: University of Oklahoma Press.

National Geographic (2011) *Guide to the National Parks of Canada.* Washington, DC: National Geographic Society.

National Institute of Anthropology and History (no date) 'Introduction'. Online: www.inah.gob.mx/index.php/academia/bibliotecas-inah (accessed 1 March 2011).

Notzke, C. (2006) *"The Stranger, The Native and the Land": Perspectives on Aboriginal Tourism.* Concord, Canada: Cactus Press.

Parks Canada (2010) *Tongait KakKasuangita SilkKijapvinga Torngat Mountains National Park of Canada. Management Plan.* Ottawa, ON: Parks Canada.

Powell, B.D. (2006) 'The memorials on Beechey Island, Nunavut, Canada: An historical and pictorial survey', *Polar Record* 42(223): 325–333.

Rondeau, R.M. (2010) 'The wrecks of Franklin's ships Erebus and Terror; their likely location and the cause of failure of previous search expeditions', *The Journal of the Hakluyt Society*. Online: www.hakluyt.com/PDF/Rondeau_Franklin.pdf (accessed 11 August 2011).

Rydell, R.W. (1984) *All the World's a Fair: Visions of empire at American international expositions, 1876–1916.* Chicago: University of Chicago Press.

Sandlos, J. (2008) 'Not wanted in the boundary: The EXpulsion of the Keeseekoowenin Ojibway Band from Riding Mountain National Park', *Canadian Historical Review* 89(2): 189–221.

Silverman, H. (2002) 'Touring ancient times: The present and presented past in contemporary Peru', *American Anthropologist* 104(3): 881–902.

Smith, V. (1977). *Hosts and Guests: The Anthropology of Tourism.* Philadelphia: University of Pennsylvania Press.

Smith, V. (1989) *Hosts and Guests: The Anthropology of Tourism*, 2nd edn. Philadelphia: University of Pennsylvania Press.

Spence, M.D. (1999) *Dispossessing the Wilderness: Indian Removal and the Making of the National Parks.* Oxford: Oxford University Press.

Stewart, E.J., Draper, D., and Dawson, J. (2010) 'Monitoring patterns of cruise tourism across Arctic Canada', in M. Luck, P. Maher, P. and E.J. Stewart (eds.) *Cruise Tourism in Polar Regions: Promoting Environmental and Social Sustainability?* London: Earthscan, pp. 147–160.

Stopp, M.P. (2009) 'Eighteenth century Labrador Inuit in England', *Arctic* 62(1): 45–64.

Veitch, A.M., and Harrington, F.H. (1996) 'Brown bears, black bears, and humans in northern Labrador: An historical perspective and outlook to the future', *Journal of Wildlife Research* 1(3): 245–250.

Wheeler, S. (2011) *The Magnetic North: Notes From the Arctic Circle*. New York: Farrar, Straus & Giroux.

Wood, R.E. (1984) 'Ethnic tourism, the state, and cultural change in Southeast Asia', *Annals of Tourism Research* 11(3): 353–374.

Zeppel, H. (1997) 'Headhunters and longhouse adventure: Marketing of Iban culture in Sarawak, Borneo', in M. Oppermann (ed.) *Pacific Rim Tourism*. Wallingford: CAB International, pp. 82–93.

# 13 Seeing it off

## Last chance tourism opportunity shagged by rare parrot!

*Eric J. Shelton*

> 'Do you know about the kakapo . . . It's the world's largest, fattest and least-able-to-fly parrot. It lives in New Zealand. It is the strangest bird I know of and will probably be as famous as the dodo if it goes extinct.'
>
> 'How many are there?'
>
> 'Forty and falling.'
>
> (Adams and Carwardine 1990: 7)

Adams and Carwardine's (1990) book *Last Chance to See*, a blend of serious conservation messages, humorous travelogue and ecocriticism, was very much of its time. It is always prudent to remember that ecocriticism, that is, writing about the natural world, and the critical reception of that writing, remains first and foremost an act of writing, firmly rooted in language (Morton 2007). The better the writing, the more seduced is the reader by the writing. *Last Chance to See* was such a good read in 1990 that some of its idiosyncrasies are obvious only today. The conceit at the heart of the text is that of Carwardine, a zoologist, explaining about critically endangered species to Adams, in this case with Adams adopting the persona of a scientifically ignorant novel writer, although already famous for *The Hitchhiker's Guide to the Galaxy* (Adams 1979). This allows Adams then to write in the first person and educate us, an uninformed but curious and sympathetic public. At the end of the book, Carwardine writes as himself to provide some closing remarks speculating how the featured species may fare. We now know it fares well. The book's topical humour, even then already rather dated had, by 2010, become anachronistic. Opaque allusions to David Lean's (1962) film version of *Lawrence of Arabia* and *Zen and the Art of Motorcycle Maintenance* (Pirsig 1974), legitimately assumed still to be decipherable to readers in 1990, cannot be assumed still to be so in 2010. Adams and Carwardine's (1990) text deserves an ongoing readership. In this chapter I argue that the story of the kakapo (*Strigops habroptilus*), rather than being an example of *last chance to see*, may better be presented as having coincided with a contemporary popular preservationist sentiment in the notion of conservation in Aotearoa/New Zealand. This is not to criticize Adams and Carwardine or to claim that in some way they misled their readers, but only to appreciate that interesting, engaging and urgent nature writing runs the risk of being an end in itself, when shorn of explanatory context.

By Carwardine's subsequent New Zealand visit with Stephen Fry (BBC 2009; Carwardine 2009, 2010), the pivotal but far-from-certain historical moment that gave urgency to their concern for the kakapo clearly had passed. Kakapo numbers had increased to over 120, from a 1989 estimate of 40, of which only 11 birds were females, due to the efforts of a dedicated Kakapo Recovery Team (Ballance 2010) (Figure 13.1).

This development of a single-species-focused, well-resourced recovery plan, detailing specific governance structures, management procedures and reporting

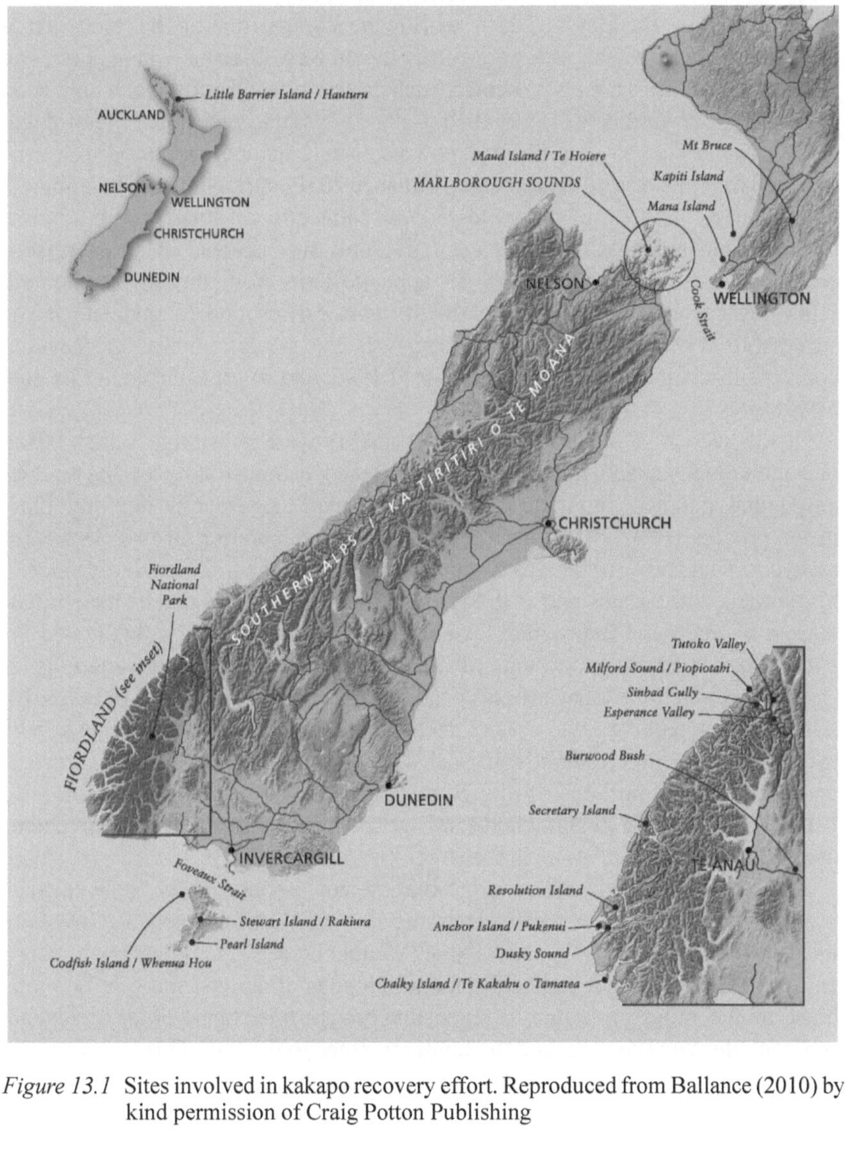


*Figure 13.1* Sites involved in kakapo recovery effort. Reproduced from Ballance (2010) by kind permission of Craig Potton Publishing

regimes associated with recovery effort, came to be used with other species; for example, the Yellow-eyed penguin (*Megadyptes antipodes*). This approach to conservation now has developed to a point of sophistication that, in 1990, had seemed unlikely ever to occur. In part, New Zealand has embraced preservationist conservation so wholeheartedly as a response to what in hindsight is viewed as a moral failing in not acting earlier to protect the endemic avian fauna (Young 2004). High rates of endemism is a feature of islands, and the introduction and effects of exotic species a common issue. New Zealand experienced particularly high rates of extinctions and has responded by establishing a network of protected areas specializing in species restoration (Sanctuaries of New Zealand 2011). With respect to Buckley's (2010) worldwide comparison of the provision of protected areas by region, it is noteworthy that in New Zealand, where about one-third of the country's area is protected, such protection uniformly is at a high level. Adams and Carwardine's (1990) at-first-glance dystopic ecocritical rhetoric with respect to the kakapo may be read now as always having supported a positive 'rescued from the brink of extinction' (Ballance 2010) narrative; one incompatible with the negative tenor of the professed *last chance to see* standpoint but hidden in plain sight amongst the language. The book may be read as an exercise in *consciousness raising*, to borrow an apposite term from the environmentally informing second-wave feminism of 20 years earlier, and to that end it was successful. Where consciousness raising was less successful was in providing context sufficient to explain the *zeitgeist* of 1990, and its crucial role in this good news story.

On this second visit, while Carwardine and Fry were revisiting Codfish Island, one individual kakapo, Sirocco, in a misplaced demonstration of the species' new-found potency, mounted Carwardine's head, vigorously, drawing blood in the process (BBC 2010); a rather undignified celebration of two decades of recovery. That Carwardine was Sirocco's target is all the more comedic since, in his pairings with Adams and with Fry both, the zoologist has played the straight-man to his partners' funny-men. Twenty years on then, and the threatened *last chance to see* tourism opportunity of 1990 had successfully been thwarted; its last chance rationale, like Carwardine's head, 'shagged by a rare parrot' (BBC 2010). Also, without needing to go extinct, the kakapo had challenged the dodo's fame: 'Somehow, this up-close-and-personal sequence found its way onto YouTube and has since had millions of hits. Sirocco became the official "spokesbird" for conservation in New Zealand and I still get people making gyrating movements and shouting "kakapo!"at me in the street' (Carwardine 2010: 10).

Thus, the kakapo and conservation have entered the connectivity age by utilizing what is by now unexceptional electronic communications capacity, involving means of seeing, previously unavailable (Walther *et al.* 2011). Adams would have enjoyed the sequence of technological productions of culture: radio programme, book of the radio programme, television programme, book of the television program, the YouTube clip and the book reference to the clip. This technology of dissemination is at the heart of the story of the kakapo renaissance.

## Background

Adams and Carwardine (1990), in their chapter on the kakapo, and in the book of the radio series (Carwardine 2010), tell of how their account of visiting Codfish Island, the centre of conservation effort for this bird, most likely assisted in its recovery:

> [W]e tell Don Merton that we think we've been forgiven. 'Oh yes, I think so,' he says. 'You may have bumbled around a bit and trodden on a few toes, but you've actually stirred things up a bit as well. The press conference was very effective, and from what I hear there's an imminent decision coming from quite high up to move the kakapo conservation programme to the top of the Department's priority list, which should mean that we get allocated more resources. I just hope it's not all too late'.
>
> (Adams and Carwardine 1990: 134)

When the pair arrived back in England, this hope seemed to be well founded:

> But the best news of all was still to come. Just before going to press, a very excited Don Merton telephoned to say that a newly made kakapo nest had been found on Little Barrier Island. Inside the nest . . . is a single kakapo egg. Transferring kakapos (sic) to Little Barrier and Codfish Islands has been a calculated risk – but it is the only hope of saving the kakapo from extinction.
>
> (Adams and Carwardine 1990: 202)

New Zealand's conservation story, within which this exchange needs to be situated, can be told comprising two parts. The first part of the account may be formulated as a narrative of declension, despoliation and despair while the second part presents a narrative of hope. This is not to claim that the narrative of hope necessarily is evidence-based; simply that it is now a powerful force within New Zealand society and increasingly is applied to endangered endemic birds besides the kakapo, including the kokako (*Callaeas cinerea*), the Yellow-eyed penguin (*Megadyptes antipodes*) and the kiwi (*Apteyrx sp.*). The despoliation narrative (Shelton and Tucker 2008: 198–99), as usually delivered, tells of how the native and endemic species, labelled by Gibbs (2006) as 'ghosts of Gondwana', suffered from the impacts of human settlement. This overwhelmingly negative and dispiriting account contains some particularly poignant moments. One was when Andreas Reischek, a late nineteenth-century ornithologist, realized that many species of New Zealand's unique avifauna were facing extinction, so deliberately shot many specimens in order to supply European museums with skins, knowing that soon there would be no more individuals left to collect and that his actions were hastening that time (Wilson 2004; Young 2004). Reischek was only one such naturalist; Audubon in the United States would have understood the logic (Graham 1990). Another poignant moment was when Richard Henry realized that the sanctuary of Resolution Island, established in 1891 with the intention of being available for the relocation of vulnerable ground-dwelling birds, had been infiltrated by weasels or stoats (*Mustela erminea*), their major introduced predator (Ballance 2010). Finally, the 1960s saw the extinction of the Stewart Island snipe (*Coenocorypha iredalei*) and the Greater Short-tailed bat (*Mystacina robusta*); events that suddenly seemed to be incongruous and unacceptable in a country that saw itself as a well-educated, Western, affluent, modern liberal democracy that prided itself on its wildlife to the extent that the kiwi was a symbol of national identity. These events occurred at a time before the notion, in this country at least, of last chance to see tourism; only scientists directly bore witness to these modern extinctions. Spurred to action by the realization that rats alone could cause island-wide extinctions, an embryonic public conservation ethic valorized Don Merton's almost single-handed rescue of the Chatham Island (or Black) robin (*Petroica traversi*) (Ballance 2010; Young 2004). Suddenly, what was of interest was an implicit new byline: *possible future opportunity to see*, rather than any attempt to develop an outmoded *last chance to see*. The exchange quoted above, from Adams and Carwardine (1990), describing the triumph of hope over despair, is as much constituted by that change as it is constitutive of it.

In the post-World War II period, often the technology of experiencing what later would be called nature conservation was through viewing the nature documentary. Such documentaries were likely to have been produced by the National Film Unit, a government department, and screened at the cinema as a short before the feature film. An example of such a film is *Pictorial Parade No.22*, of 1954 (Archives New Zealand 2010), which contains a segment titled *Possum in the Bush*. Despite the narrator's jaunty delivery of a script, typical of its time, and demonstrating an unbounded faith in progress, still the underlying narrative is one of despoliation. The clip describes endemic and native flora suffering depredation from this introduced browser (*Trichosurus vulpecula*) and the catastrophic implications if the situation were allowed to continue unchecked. In the Aotearoa/New Zealand imaginary one seeks recreation and solace in 'the bush' (Ross 2008). Quite clearly, the message here was that 'the bush' was under threat but human development of some sort would save it. In fact, it was human development of the economic kind that proved to be the largest threat.

## Context for saving the Kakapo: saving Manapouri

Hydroelectric power has long been the mainstay of New Zealand's electricity production, providing approximately 60 per cent of the nation's supply (New Zealand Ministry for Economic Development 2006), and has involved the damming of many rivers. This process had continued for decades, without significant protest, or in fact without the availability of any effective mechanism for protest. New dams meant that from about 1960 electricity could be provided without interruption, to most households, and 'electric hot water' could replace the use of wetback coal ranges. A proposal to raise the levels of lakes Te Anau and Manapouri by up to 30 metres, thus creating one lake, proved to be a turning point in this acceptance of the inevitability and desirability of progress (Peat 1994; Young 2004).

Not only would raising the lake levels inundate large areas of bush-clad shoreline but also this time the purpose of the electricity generation was not to be for domestic use but to supply cheap energy to a proposed foreign-owned aluminium smelter. In response to extensive nationwide protest the centre-left Labour party electioneered on limiting the lake level fluctuation and, after being elected in 1972, set these limits. This rescue of the natural environment from the forces of international capital set the scene for subsequent, though not always successful, opposition to large logging and hydro schemes. In particular, with respect to the idea of narratives of hope, the Save Manapouri campaign provided a narrative position that now could be presented either in public or in private, and that would be understood, even by its opponents, since it had been so extensively articulated in the popular media. The dispute over Manapouri has remained a touchstone for groups seeking environmental protection in other settings; an ongoing symbol of good versus evil, even when the issues involved have been dissimilar. Hundreds of thousands of visitors every year travel along the lakefront on their way to Milford Sound. Nearby Lake Monowai, inundated for a hydro-electricity scheme in 1925, provides a salutary example of what a lake whose shoreline was previously fine-gravelled beaches looks like still, 80 years after the verdant lakefront forest was flooded (Liefting 2008) but '[t]he lessons about lake raising of lake Monowai seemed to have made no impression on the engineers and planners' (Peat 1994: 22). This clash of aesthetics – Manapouri as beautiful and Monowai as ugly – was central to the dispute and was encapsulated in a song that was used during the campaign:

> Damn the dam cried the fantail,
> As he flew into as he flew into the sky,
> To give power to the people
> All this beauty has to die.
>
> (Hanlon 1973: chorus to *Damn the Dam*)

The power of this lyric, sufficient to have it banned from publically funded radio (Peat 1994: 57), lay in how easily it simultaneously evoked, and could be situated within, a despoliation narrative linking aesthetic loss, destruction of 'the bush' and therefore of national identity, loss of habitat and the extinction of species, even though, in fact, the project carried no likelihood of any extinctions. Of at least equal importance was the rejection of the narrative of 'progress at almost any cost'. The scheme was identified by its opponents as an example of economic boosterism. The aluminium company for whom the power project was being constructed was Swiss and its negotiators had managed successfully to buy the power for less than it cost the New Zealand taxpayer to produce it. Suddenly, progress was not such a simple concept from which everyone benefitted, it was now obvious that there were winners and losers. The point here is that the idea of labelling some feature of the natural environment, be it a lake or a wild river or a species of wildlife, as being the last one available to visit or view, and somehow to celebrate that status, had rapidly become inconceivable. Scarcity coupled with the likelihood of something disappearing altogether was not to become an acceptable point of difference for marketing the viewing of any aspect of the natural world; preservation and conservation, leading to an abundance of individuals and species was to be the new focus.

The Department of Conservation (2010a), formed in 1987, began identifying exemplary ecosystems, landforms, landscapes and habitats specifically to protect them within what in Aotearoa/New Zealand is called the conservation estate. New Zealand is a country with a population of around 4.4 million persons and of its landmass of 268,676 $km^2$ (Statistics New Zealand 2010) approximately one-third now enjoys some form of protection above and beyond that offered by the comprehensive 1991 Resource Management Act (New Zealand Ministry for the Environment 1991), the key piece of legislation governing planning and development and which has operated more or less successfully for 20 years with only minor amendments. When Adams and Carwardine visited in 1990, this production and management of protection, on this scale, was in its infancy. The forced merging of part of the Forest Service, Wildlife Division of the Department of Internal Affairs, and Lands and Survey Department in order to create the Department of Conservation had been fraught with difficulty; individuals with deeply held but also deeply incompatible environmental views, particularly around the virtues of development of land, sometimes were expected to share an office in the new organization (University of Waikato Conserv-Vision Conference 2007 pers. comm.).

Also, ever since the extensive economic restructuring of 1984, conservation in New Zealand has been enmeshed with the resulting neoliberal economic ideology coupled with an already existing Romantic aesthetics of nature. These relationships, through the use of tourism as a driver of economic growth, have made endangered species' protection inseparable from the functioning of the late capitalist economic system. *Last chance to see* tourism product could easily be situated within the same set of economic and aesthetic relations already operating, that is, global capitalism and the viewing of charismatic megafauna. Adams and Carwardine's (1990) boost to kakapo protection occurred at a time when the post-Manapouri purely protectionist conservation ethic suddenly was being critiqued and challenged by the government responsible for it. Various categories of protection were gazetted, from Scenic Reserve at one end of the spectrum to Wilderness Area at the other, and each allowed a range of activities to be pursued (Department of Conservation 2010b). Nevertheless, 'it would be naive to believe that the present (Conservation) estate boundaries are somehow sacrosanct, or that some future government may not allow extractive industries to operate within the estate' (Shelton and Tucker 2008: 203).

This statement proved prescient when, in 2009, Gerry Brownlee, Minister of Economic Development, proposed removing from protection conservation land then exempt from mining, under Schedule Four of the Crown Minerals Act (Haggart 2010). This proposal was withdrawn on 20 June 2010. 'Mr. Brownlee said that of 37,000 submissions on the issue, most said no land should be removed from schedule four. "We heard that message loud and clear," he said' (NZPA 2010: 2). 'Environmental Defense Society chairman Gary Taylor said mining would have clashed with New Zealand's "clean green brand and image"' (Bennett 2010: 1), a quote best situated within a New Zealand imaginary. To date, tourism development largely has avoided such criticism. These actions, linking conservation and economics, made explicit the process that had been in train implicitly since DOCs' formation. DOC's function was, at its inception and has ever since been, a mixture of conserving and making available, fostering recreation and allowing tourism on conservation land, providing the use is consistent with the conservation of the resource (Department of Conservation 2011). This recent overt championing of the conservation economy makes it now unavoidable "to recognize that capitalist policies and values, and often neoliberal policies and values, pervade conservation practice" (Brockington, Duffy and Igoe 2008: 3). The New Zealand Institute of Economic Research (NZIER) presented 'Insight 15: Conservation and Mining', which stated:

> If more funding were available, the additional revenue could be used to acquire new lands of greater priority for protection and to shift the portfolio of conservation lands away from the current preponderance of mountain, rock and forest to include more of the scarcer and more vulnerable lowland habitats.
>
> To do this, however, requires moving away from the notion that once land is acquired for conservation it is closed for all future development other than the most low impact tourism or recreation uses that are deemed compatible with conservation. That approach ossifies the conservation estate and is ultimately unsustainable, both economically and environmentally. For as climate changes, whether through human or natural causes, species range and habitats shift, and the pattern of national parks and reserves will need to be reconfigured if it is not to become functionally obsolete for its conservation purpose.
>
> (NZIER 2010: 3)

This idea that conservation land may be redesignated to allow other than low-impact activity, and particularly to allow mining, proved anathematic to prominent conservation organizations; for example, The New Zealand Forest and Bird Protection Society (Wards 2010) and groups who recreate within conservation areas, including the Federated Mountain Clubs of New Zealand (Barnett 2009). Rightly or wrongly, an emphasis on economic considerations in the processes of the creation, management, retention or disestablishment of protected areas, and in particular the application of neoliberal market models to these processes, is perceived by these conservation organizations as putting the whole notion of conservation at risk; the opposite of the NZIER position. Central to any conception of conservation is this notion of protection, whether achieved through management informed by a preservationist ideology or neoliberal market-based ideology. If we accept the proposal that conservation in the twenty-first century is well and truly enmeshed with neoliberalism, 'speaking comfortable truths to power'

(Brockington *et al.* 2008: 3), then it is inescapable that protection be viewed as part of the late-capitalist economic production and consumption system; the production of protection then becomes an overtly economic and political act every bit as much as it involves conservation effort being expended to produce conservation outcomes.

## The contemporary scene: no room for pessimism

An important aspect of the success of the Save Manapouri campaign was the appointment by central government of the Guardians of Manapouri, a novel arrangement comprising a small group of individuals with powers and responsibilities of governance but who are not, in contrast to the Department of Conservation, public servants employed by the government of the day and forbidden to make public statements contradictory to government policy (Peat 1994). The appointment of guardians ushered in a new form of environmental governance (Fletcher 2010) that has proved to be particularly suited to individual species recovery planning. Ruhanen, Scott, Ritchie and Tkaczynski (2010: 6) identified 'three common characteristics, or generally agreed principles of governance . . . [g]overnance is much broader than government . . . governance implies less government control . . . [and] governance involves multiple stakeholders who have an interest in the specified task'.

In New Zealand, endangered-species recovery success stories commonly now have three collaborative elements of governance: the Department of Conservation as the regulatory authority with overall statutory responsibility for species management, an environmental non-governmental organization (ENGO), often in the form of a not-for-profit Incorporated Society, and corporate sponsorship of some sort. This structure, especially the corporate sponsorship element, militates against the public presentation of narratives of declension – for example, the situating of a species recovery programme within the sixth major extinction event, as Carwardine does (in Adams and Carwardine 1990: 204): 'As zoologists and botanists explore new areas, scrabbling to record the mere existence of species before they become extinct, it is like someone hurrying through a burning library desperately trying to jot down some of the titles of books that will now never be read.' (See Eldredge (2010) for a web-based scientific educational resource on this topic.) Sponsors are interested in winners and good news, which leads to a focus on a reporting of the small scale, the immediate and the successful. Also, sponsors are mindful of their need for balance-sheet outcomes as well as for conservation outcomes (Hacking and Guthrie 2008). By formulating species' recovery challenges at the species level and not at a whole-of-ecosystem level, or at the level of climate change or mass extinction, conservation-minded individuals may more easily believe that they have some agency with respect to halting biodiversity loss (Clayton and Myers 2009). When whole-of-ecosystem species-recovery effort involves translocations to small offshore islands, as is the case, for example, on Tiritiri Matangi (Supporters of Tiritiri Matangi Open Sanctuary 2010), there may be difficulties in establishing and maintaining sustainable biotic assemblages. In their new island setting, some translocated species have proved to be incompatible with others. Naturally occurring revegetation may threaten to become dominated by a single plant species (e.g. *Metrosideros sp.*) and a mosaic of more varied types may need to be established and actively managed. The situation on Codfish Island, itself a small island, is that kakapo are privileged with respect to management of the island and the primary focus is on their requirements. Richard Henry's 1890s dream of nearby Resolution Island as a sanctuary is now being realized and it is large enough to accommodate a whole-of-ecosystem approach (Department of Conservation 2008). Currently, as well as fully protected offshore islands, there is a growing network of fenced and unfenced mainland sanctuaries (Shelton and Tucker 2008), either completely without introduced browsers and predators or with them in very low densities (Sanctuaries of New Zealand 2010). The mode of presentation of this conservation effort, thanks to the development of new technologies of dissemination and connectivity mentioned above, has changed from the nature documentary to the interactive website (BNZ Save the Kiwi 2010; Department of Conservation 2010c; Kaharoa Kokako Trust 2010; Kakapo Recovery 2010; Yellow-eyed Penguin Trust 2010). Each of these bodies has corporate support closely involved, directly or indirectly, with their governance. The Bank of New Zealand (BNZ) supports Kiwi conservation, Blakely Pacific (forestry) supports Kokako projects, RioTintoAlcan (aluminium smelting) supports kakapo work and Fonterra (dairying, through *Mainland* brand, support Yellow-eyed penguin recovery). This form of governance, unthinkable before Manapouri, is now so well integrated into conservation that the Department of Conservation Conservancy Newsletters routinely report on their partnership activities (Department of Conservation 2010d). Alison Ballance's (2010) handsome coffee-table format book, *Kakapo: Rescued from the Brink of Extinction*, was produced by *Kakapo Recovery* to celebrate 20 years of Department of Conservation-Forest and Bird-RioTintoAlcan collaboration – copies have been donated to all intermediate and secondary schools in the country. This focus on successfully providing effective protection to a growing list of charismatic birds raises important issues of resourcing. The forest floor is rich in interesting endemic invertebrates and is a major contributor to ecosystem resilience (Young 2004) yet receives little press and less sponsorship. Historically, hundreds of millions of sea birds roosted, nested and died in the forest, transporting essential nutrients from the sea to the land (Wilson 2004). For effective ecosystem restoration long term, the forest floor and the nutrient cycle from the sea to the forest must be re-established. Even within sanctuaries, remedying this chronic nutrient deficit may well be a bigger challenge than single-species recovery.

As for Sirocco, he long ago inadvertently imprinted on humans and consequently cannot take part directly in the breeding programme, finding other kakapo unappealing. Instead, Sirocco, when not on Codfish Island, occasionally has been made available for public viewing, including for a while featuring in night tours in a special building at Auckland Zoo, where he attracted large crowds. Carwardine's mounted head provides an embodied and enduring, if displaced, metaphor for a triumph of Eros over Thanatos. Where conservation has embraced neoliberal global capitalism, then *last chance to see* is morbid; *rescued from the brink of extinction* is inspiring. The urgency and magnitude of conservation effort required in the face of the threatened next round of mass extinction has no place for a tourism of morbidity. In New Zealand at least, *last chance to see* has been well and truly shagged by a rare parrot.

## References


Adams, D. (1979) *The Hitchhiker's Guide to the Galaxy*, London: Pan Books.

Adams, D. and Carwardine, M. (1990) *Last Chance to See*, London: Heinemann.

Archives New Zealand (2010) 'Pictorial Parade No. 22'. Online: http://audiovisual.archives.govt.nz/wiki/index.php/PICTORIAL_PARADE_No._22 (accessed 3 November 2010).

Ballance, A. (2010) *Kakapo: Rescued from the brink of extinction*, Nelson: Craig Potton Publishing.

Barnett, S. (2009) *Editorial: The Mines of Moria*, FMC Bulletin. Wellington: Federated Mountain Clubs of New Zealand.

BBC (2009) 'Shagged by a rare parrot – Last Chance To See – BBC Two'. Online: www.youtube.com/watch?v=9T1vfsHYiKY (accessed 1 November 2010).

BBC (2010) 'Last Chance to See'. Online: www.bbc.co.uk/lastchancetosee/ (accessed 1 November 2010).

Bennett, A. (2010) 'Mining plans scrapped – national parks to be protected', *The New Zealand Herald*, 20 July.

BNZ (2010) 'Save the Kiwi'. Online: www.kiwirecovery.org.nz/ (accessed 1 November 2010).

Brockington, D., Duffy, R. and Igoe, J. (2008) *Nature Unbound: Conservation, capitalism and the future of protected areas*, London: Earthscan.

Buckley, R. (2010) *Conservation Tourism*, Wallingford: CAB International.

Carwardine, M. (2009) *Last Chance to See: In the footsteps of Douglas Adams*, London: Collins.

Carwardine, M. (2010) 'Foreword', in A. Ballance (ed.) *Kakapo: Rescued from the brink of extinction*, Nelson: Craig Potton Publishing.

Clayton, S. and Myers, G. (2009) *Conservation Psychology: Understanding and promoting human care for nature*, Chichester: Wiley-Blackwell.

Department of Conservation (2008) 'Important restoration project for Resolution Island'. Online: www.doc.govt.nz/about-doc/news/media-releases/2008/important-restoration-project-for-resolution-island/ (accessed 1 November 2010).

Department of Conservation (2010a) 'The Conservation Act 1987'. Online: www.doc.govt.nz/about-doc/role/legislation/conservation-act/ (accessed 1 November 2010).

Department of Conservation (2010b) 'Places to Visit'. Online: www.doc.govt.nz/parks-and-recreation/places-to-visit/ (accessed 1 November 2010).

Department of Conservation (2010c) 'Then and Now'. Online: www.kakaporecovery.org.nz/index.php?option=com_content&view=section&id=9&Itemid=237 (accessed 1 November 2010).

Department of Conservation (2010d) *Good as Gold: Penguin partnership on Otago Peninsula*, Otago Conservancy Newsletter, 1.

Department of Conservation (2011) 'A short history of the Department of Conservation 1987–2007: Born with a mission'. Online: www.doc.govt.nz/publications/about-doc/a-short-history-of-doc/born-with-a-mission/ (accessed 28 February 2011).

Eldredge, N. (2010) 'The Sixth Extinction'. Online: www.actionbioscience.org/newfrontiers/eldredge2.html (accessed 1 November 2010).

Fletcher, R. (2010) 'Neoliberal environmetality: Towards a poststructuralist political ecology of the conservation debate', *Conservation and Society* 8: 171–81.

Gibbs, G. (2006) *Ghosts of Gondwana: The history of life in New Zealand*, Nelson: Craig Potton Publishing.

Graham, F. Jr. (1990) *The Audubon Ark: A History of the National Audubon Society*. New York: Alfred A. Knopf.

Hacking, T. and Guthrie, P. (2008) 'A framework for clarifying the meaning of triple bottom-line, integrated, and sustainability assessment', *Environmental Impact Assessment Review* 28: 73–89.

Haggart, M. (2010) 'More conservation areas a possibility', *Otago Daily Times*, 6 January, p. 13.

Hanlon, J. (1973) *Damn the Dam*. AHI.

Kaharoa Kokako Trust (2010) Online: www.kokako.org.nz/ (accessed 1 November 2010).

Kakapo Recovery (2010) Online: www.kakaporecovery.org.nz/ (accessed 1 November 2010).

Liefting, A. (2008) 'Stumps from drowned trees at the north eastern end of Lake Monowai', Online: www.google.co.nz/imgres?imgurl=http://upload.wikimedia.org/wikipedia/commons/b/bd/LakeMonowai__drowned_tree_stumps.jpg&imgrefurl=http://commons.wikimedia.org/wiki/File:Lake_Monowai_drowned_tree_stumps.jpg&usg=__iOwoKWWd43NDiFpqouSTM8e9XXI=&h=1694&w=2814&sz=842&hl=en&start=12&sig2=O9SEqGx6tUJ0bipOeMc5lQ&zoom=1&um=1&itbs=1&tbnid=Xg8XacOOBIWPbM:&tbnh=90&tbnw=150&prev=/images per cent3Fq per cent3Dlake per cent2Bmonowai per cent2Bnz per cent26um per cent3D1 per cent26hl per cent3Den per cent26safe per cent3Doff per cent26sa per cent3DN per cent26tbs per cent3Disch:1&ei=jUL0TJiwOZPKcODxmlgK (accessed 1 November 2010).

Morton, T. (2007) *Ecology – Without Nature: Rethinking Environmental Aesthetics*, Massachusetts: Harvard University Press.

New Zealand Institute for Economic Research (2010) 'Conservation and mining', NZIER *Insight* 15. Online: http://nzier.org.nz/ (accessed 1 November 2010).

New Zealand Ministry for the Environment (1991) 'Resource Management Act'. Online: www.mfe.govt.nz/rma/index.html (accessed 28 February 2011).

New Zealand Ministry of Economic Development (2006) 'Electricity market review'. Online: www.med.govt.nz/templates/ContentTopicSummary 24877.aspx (accessed 28 February 2011).

NZPA (2010) 'Conservationists rejoice – concerns remain about future mining of Crown Land', *Otago Daily Times*, 21 July.

Peat, N. (1994) *Manapouri Saved! New Zealand's first great conservation success story*, Dunedin: Longacre Press.

Pirsig, R. (1974) *Zen and the Art of Motorcycle Maintenance*, New York: William Morrow & Company.

Ross, K. (2008) *Going Bush: New Zealanders and Nature in the Twentieth Century*, Auckland: Auckland University Press.

Ruhanen, L., Scott, N., Ritchie, B. and Tkaczynski, A. (2010) 'Governance: A review and synthesis of the literature', *Tourism Review* 65: 4–16.

Sanctuaries of New Zealand (2010) Online: www.sanctuariesnz.org/ (accessed 1 November 2010).

Shelton, E.J. and Tucker, H. (2008) 'Managed to be wild: Species recovery, island restoration, and nature-based tourism in New Zealand', *Tourism Review International* 11: 197–204.

Statistics New Zealand (2010) Online: www.stats.govt.nz/ (accessed 1 November 2010).

Supporters of Tiritiri Matangi Open Sanctuary (INC) (2010) 'Tiritiri Matangi Open Sanctuary'., Online: www.tiritirimatangi.org.nz/ (accessed 1 November 2010).

Walther, J., Carr, C., Choi, S., Deandrea, D., Kim, J., Tong, T. and Van der Heide, B. (2011) 'Interaction of interpersonal, peer, and media influence sources online: A research agenda for technology convergence', in Papacharissi, Z. (ed.) *A Networked Self: Identity, community, and culture on social network sites*, New York: Routledge.

Wards, B. (2010) 'Editorial', *Forest & Bird*, Wellington: Kalamazoo Wyatt & Wilson (NZ) Ltd.

Wilson, K.J. (2004) *Flight of the Huia: Ecology and conservation of New Zealand's frogs, reptiles, birds and mammals*, Christchurch, Canterbury University Press.

Yellow-eyed Penguin Trust (2010) Online: http://yellow-eyedpenguin.org.nz/ (accessed 1 November 2010).

Young, D. (2004) *Our Islands, Our Selves: A history of conservation in New Zealand*, Dunedin, University of Otago Press.

# 14 Last chance tourism in Canada's protected areas

## Management implications and emerging ethical considerations

*Chris Lemieux and Paul Eagles*

### Introduction

First established to preserve scenic wonders and tourist attractions, in recent years the rationale for protected areas has changed to become a cornerstone of biodiversity conservation and ecological sustainability, as well as an important barometer of world ecosystem health and the human condition (United Nations 2005). According to the World Conservation Union (IUCN) and the United Nations Environment Programme's World Conservation Monitoring Centre (WCMC), the world's 17,500+ parks and other forms of protected areas cover approximately 11.6 percent of Earth's terrestrial base (IUCN and UNEP-WCMC 2011). In Canada, 5,300+ protected areas cover nearly 10 percent of the nation's terrestrial base and represent the third largest land-use type behind forested areas and permanent pasture lands (Canadian Council on Ecological Areas 2011). Despite the recent volatility of economic markets worldwide, protected areas continue to attract a high number of visitors and contribute to a significant amount of tourism-related economic activity in nearby communities. Canadian national parks alone attract more than 13 million visits per year and personal benefits obtained from visitation are the key element in societal acceptance and the approval of parks and their management (Bushell and Eagles 2007; Parks Canada 2009).

Many ecotourism and last chance tourism attractions are located within the world's protected areas estate. Examples include rare or endangered species and/or ecosystems (e.g., polar bears, coral reefs), species and ecosystems whose distributions are highly influenced by climate (e.g., migratory species, such as birds and other wild life), physical features (e.g., glaciers), and commemorative resources (e.g., historic sites). In fact, protected areas are often legislated specifically to permanently protect, restore, and provide opportunities to appreciate these resources (e.g., Government of Canada 2000; Government of Canada 2002; Government of Ontario 2006). Furthermore, park facilities and infrastructure (e.g., boardwalks and interpretive trails) are often explicitly situated and designed to facilitate their viewing and appreciation (Figure 14.1).

Climate change will affect the resource base upon which nature-based tourism opportunities in Canada's protected areas are predicated. As discussions throughout this book have revealed, climate change has sufficiently begun to impact

*Figure 14.1* Trail marker indicating the former extent of the Athabasca Glacier, Columbia Icefield, in Jasper National Park, Alberta, Canada (Photo credit: Maggie Tacheny, 2007)

critical, once relatively stable climate regimes. The Intergovernmental Panel on Climate Change's (IPCC) *Fourth Assessment Report* (AR4) has clearly substantiated that climate change induced by human-generated greenhouse gas (GHG) emissions is now implicated in a myriad of coincident impacts: perturbations in regional temperature regimes and precipitation patterns; severe weather events; sea level rises; changes in ecosystem composition, structure, and function; and, projected extinctions of species (IPCC 2007). In ecological terms, the anticipated shifting of species distributions – consequences less commonly visible than impacts on human settlements and livelihoods – may well be the most disruptive long-term impacts. As fixed assets established to conserve representative samples of ecosystems, species, and other assets, protected areas worldwide are highly vulnerable to shifting climate regimes.

Climate change-induced ecological and physical changes to resources housed within protected areas are likely to stimulate a diversification of nature-based tourism products and destinations. Climate change could affect visitor attractions, visitation patterns, and the overall quality of visitor experiences within protected areas, often in unpredictable ways. Given these implications, it will be important to determine how to manage for climate change-induced visitor behavior responses and the shifting and emerging of destination and recreational preferences in these areas (e.g., how to adapt infrastructure, accommodate losses in destination preferences, and minimize conflicts between user groups). There will also be an overriding need to ensure visitor safety as more tourists come to experience vulnerable resources.

Last chance tourism is a tourism trend whereby tourists travel to endangered natural sites to see them before they vanish or are irrevocably transformed. Travel motives are many in number and are complex in application. However, Dawson *et al.* (2010) found evidence that some people's travel to observe polar bears in protected areas (i.e., Churchill Wildlife Management Area, Wapusk National Park) in northern Manitoba is stimulated by a desire to see this southern population of bears before climate change affects their distribution within the park (also see Lemelin *et al.* 2010). Extrapolation from this finding would lead to the concept that this tourist travel motive could become more prominent in the future as the natural environments and the species therein are threatened by climate change. Therefore last chance tourism may become a more prominent travel motive for many more parks and protected areas in this century.

This chapter provides an exploratory analysis of the implications of last chance tourism for Canada's protected areas and is divided into four sections. The first provides a discussion on climate change-induced impacts and the implications for protected areas, with particular attention paid to the implications of ecological change for park visitation. The chapter subsequently discusses management and operational challenges with respect to last chance tourism, and concludes with a discussion on some of the emerging ethical dilemmas that protected areas managers and planners will be confronted within an era of rapid climate change. While the examples provided within this chapter are exploratory in nature and focus primarily on Canada, many of the issues and ethical dilemmas detailed are also relevant to natural heritage stewards responsible for the care and protection of private lands and waters and to agencies responsible for protected areas in countries throughout the world.

## Climate change and changing park resources

Climate change has been recognized as a global concern for protected areas for nearly two decades. At the fourth World Congress on National Parks and Protected Areas, the IUCN concluded that, 'Climatic change represents a critical and urgent threat to all ecosystems . . . [and that] Existing . . . protected areas may not provide adequate future safeguards for the continued survival of existing ecosystems and species in a changing world' (WCPA 1993: n.p.). Ten years later, little progress had been made in addressing the issue. This, in addition to a growing concern over climate change in many nations, prompted a major recommendation at the fifth IUCN World Congress in Durban, South Africa (2003); Recommendation 05, 'Climate Change and Protected Areas', emphasized the need to protect biodiversity in the face of climate change through a three-fold response: (1) the mitigation of climate change by stabilizing GHG concentrations; (2) by instituting new conservation strategies, including the establishment of new protected areas that are specifically designed to be resilient to climate change-related impacts; and (3) the creation of corridors to facilitate the migration of biodiversity (WCPA 2003).

Since the fifth IUCN World Congress, the implications of climate change for protected areas policy, planning, and management has received growing attention within many nations and in the scientific literature. Conservation scientists and practitioners from several nations, including the United States (Baron *et al.* 2009; West *et al.* 2009; Joyce *et al.* 2009; Griffith *et al.* 2009; Mawdsley *et al.* 2009), Australia (Dunlop and Brown 2008), the European Union (Araújo 2009a, 2009b), Canada (Scott and Lemieux 2005; Lemieux and Scott 2005; Welch 2005; Scott and Lemieux 2007; Lemieux *et al.* 2010, 2011a, 2011b) and the U.K. (e.g., Huntley 2007; ENPAA 2009) have completed national assessments on the implications of climate change for their respective protected area systems. A few nations have since begun to develop response strategies (see NPS 2010 for example).

One of the more consistent findings within these assessments is the recognition that conservation planning based on protecting representative samples of species, ecosystems, and other resources will have to address the fundamental problem of defining what is to be considered 'representative' in an era characterized by transitory ecological and physical systems. Temperature increases and the other changes in climate that have occurred over the last 50 years are impacting the distribution and abundance of physical features and life on Earth as we know it (e.g., Parmesan and Yohe 2003; Root *et al.* 2003; Parmesan 2006; IPCC 2007). Alarmingly, climate change has been attributed to several species extinctions (Pounds *et al.* 1999; McLaughlin *et al.* 2002; Thomas *et al.* 2006), a phenomenon that will be exacerbated in the future as temperatures continue to rise and as precipitation and wind patterns change (Thomas *et al.* 2004; Schwartz *et al.* 2006; IPCC 2007).

Future climates likely will be significantly different from current climates, which will contribute to change in the composition, structure, and function of the Earth's ecosystems. Correlative 'climatic envelope' methods have been applied to species modelling to project the future impacts of climate change on terrestrial ecosystems and species. Studies consistently suggest that there will be major shifts in species and significant changes to ecosystems globally (Figure 14.2). Substantial reductions in the extent of northern species' bioclimatic envelopes and the expansion of more southerly bioclimatic envelopes have been consistently projected (e.g., Malcolm *et al.* 2002; McKenney *et al.* 2007a, 2007b; Lawler *et al.* 2009). Some projections estimate that the extent of Canada's boreal forest bioclimatic envelope could be reduced by as much as 50 per cent, with more southern areas being replaced by temperate bioclimatic envelopes (Malcolm *et al.* 2002; Gray 2005).

Overall, evidence from virtually all meta-analyses completed to date indicates that climate change will negatively impact many species in ecosystems. At the global scale, for example, Malcolm *et al.* (2006) projected that 39 to 43 per cent of biota in 'biodiversity hot spots' could face extinction under a $2xCO_2$ climate chance scenario (representing the potential loss of some 56,000 endemic plant species and 3,700 endemic vertebrate species globally). Thomas *et al.* (2004)

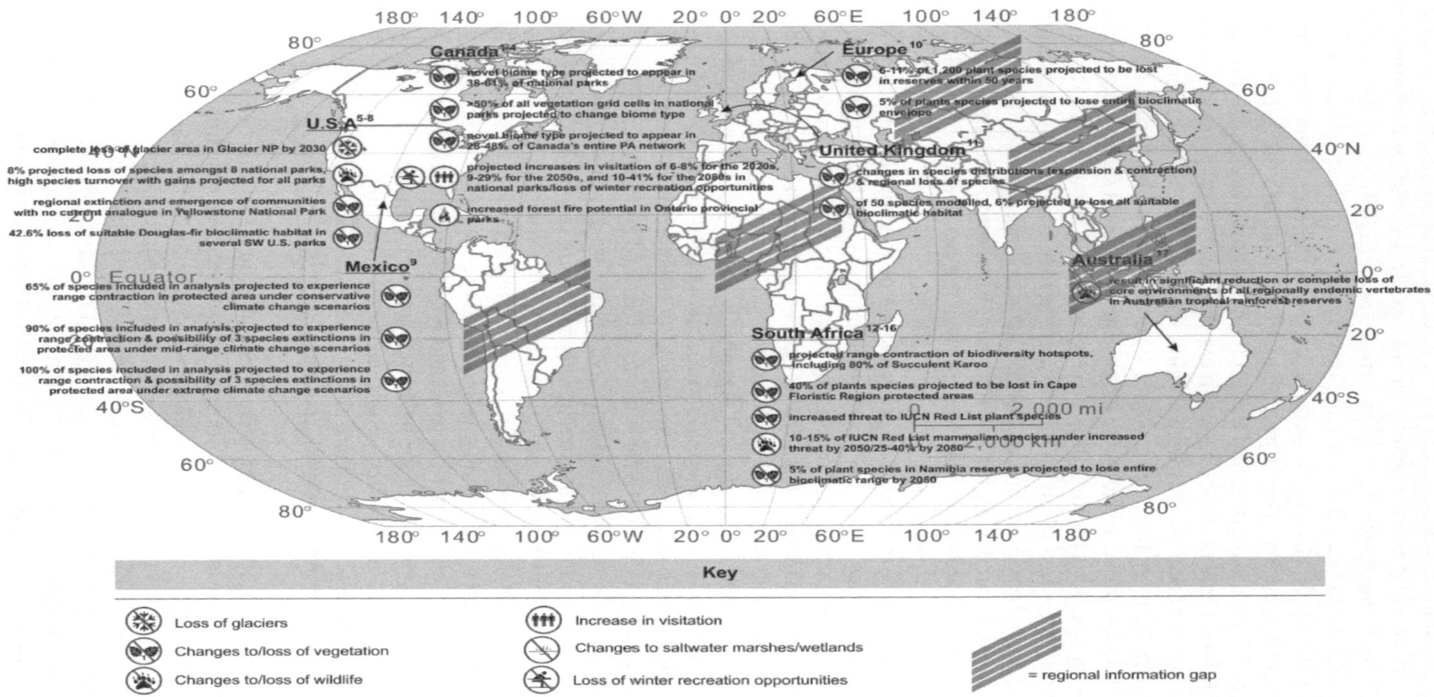


*Figure 14.2* Selected examples of projected climate change impacts in protected areas around the world. Sources: 1 = Scott *et al.* (2002); 2 = Lemieux and Scott (2005); 3 = Jones and Scott (2006a, 2006b); 4 = Hall and Fagre (2003); 5 = Burns *et al.* (2003); 6 = Bartlein (1997); 7 = Coulston and Riitters (2005); 8 = Téllez-Valdés and Davila-Aranda (2003); 9 = Rutherford *et al.* (1999); 10 = Hannah and Salm (2005); 11 = Bomhard *et al.* (2005); 12 = Thuiller *et al.* (2006a); 13 =Thuiller *et al.* (2006b); 14 =Araújo *et al.* (2004); 15 = UK Climate Impacts Programme (2001); 16 = Williams *et al.* (2003)

similarly projected that between 15 to 37 percent (depending on the climate change scenario used and the migration capacity of the species) could be committed to extinction by 2050. A more recent study by Lawler *et al.* (2009) found that 80 percent of the climate projections based on a relatively low GHG emissions scenario result in the local loss of at least 10 percent of the vertebrate fauna over much of North and South America. The largest changes in fauna are projected for the Canadian tundra, Central America, and the Andes Mountains where, assuming no dispersal constraints, some areas are likely to experience over 90 percent turnover (i.e., the replacement of one species with another species). The authors concluded that faunal distributions in the future would bear little resemblance to those of today. In a Canadian context, while range extensions will result in the introduction of new species in southern regions, this novelty may not compensate for range contractions and species loss likely to be prevalent across northern regions. Taken collectively, these studies indicate that Earth's natural landscapes and waterscapes, especially those located in alpine and Arctic regions, could undergo rapid and significant climate-induced ecological change over the next century (see Steiger *et al.* and Johnston *et al.* this volume).

Alpine glaciers housed in protected areas are an observable example of currently vulnerable resources located within protected areas. With nearly two-thirds of all national park visits annually, mountain parks are the highest-visited national parks in Canada (Parks Canada 2009). Glacier coverage in the southern Canadian Rocky Mountains is estimated to have decreased 25 percent over the twentieth century (Blais *et al.* 1998; Luckman and Kavanagh 2000). For example, the terminus of the Athabasca Glacier, the main attraction at the Columbia Icefields in Jasper National Park (Alberta), has retreated 1,200 metres since 1900 (State of the Canadian Cryosphere 2011) (refer back to Figure 14.1). Several glaciers in Auyuittuq National Park on Baffin Island (Nunavut) are currently exhibiting a steady retreat of approximately 8 meters per year (despite the Inuit interpretation of its name: 'the land that never melts'). Similarly, the Illecillawaet Glacier in Glacier National Park (British Columbia), one of Canada's most thoroughly studied glaciers, has retreated more than a kilometer since the 1880s (Parks Canada 2011a). A broad range of other inter-regional and regional climate change impacts have been identified as significant for protected areas in Canada (e.g., sea-level rise, waning permafrost, altered freshet and seasonal hydrology, wetland loses, glacial retreat, and increased insect harassment) (see Scott and Suffling 2000 and Lemieux *et al.* 2010 for reviews), all of which will have implications for last chance tourism pursuits.

While the ecological impacts of global climate change on species, ecosystems, and physical resources have become prominent subjects of research in recent years, the possible impacts of climate change on commemorative integrity (e.g., cultural heritage) have been less explored. Climate change will also have significant implications for the planning and management of cultural heritage resources such as buildings and works, in-situ archaeological remains, artifacts, and places. Floods, changing precipitation patterns, melting of permafrost, erosion, and sea-level rise, for example, could have significant effects on historic sites and on the social fabric of traditional societies (Figure 14.3, see also Lemelin and Baikie, this volume; Johnston *et al.*, this volume). Regardless of what is being protected, the extant literature on the impacts of climate change for biodiversity and other resources in Canada overwhelmingly suggests that many protected areas could become increasingly vulnerable by the end of the twenty-first century. Indeed, the increased exposure of already vulnerable resources to climate change, in combination with new and emerging vulnerable resources, could possibly render some parks and protected areas as warehouses of last chance tourism attractions.

Climate change has the potential to affect both the supply and the demand aspects of tourism. Most of the literature and most of this book chapter deals with the supply aspects: the features in the tourism destinations that attract tourists. Much less attention is given to the demand aspects, that is, the climate change impacts that will alter economies and life styles of regions and nations and thereby profoundly affect the ability of people to travel. The snow birds travel phenomenon is well known in Canada, whereby over one million Canadians travel south in winter each year to warmer climes in the Caribbean, Mexico and southern USA

*Figure 14.3* Ice scouring and permafrost instability contribute to coastal erosion at York Factory National Historic Site, Manitoba, Canada. The rate of erosion is about 3 meters/5 years and the remains of two earlier York Factories have completely disappeared. In 2008, Parks Canada Agency invested 1.75 million dollars CDN in a five-year research and engineering project aimed at finding ways to protect the national historic site. (Photo credit: Ian Beatty, www.ianbeattyphoto.com)

areas. It is possible that in the future we may see a heat birds phenomenon whereby people migrate north from the USA to Canada in the hotter summers to experience relatively cooler conditions for outdoor recreation. Other similar transformations brought about through climate changes are also expected.

## Climate change and tourism in protected areas

Parks and other forms of protected areas, such as national and provincial, have long been regarded as the cornerstones of conservation and are called for under the UN Convention of Biological Diversity (Article 8). However, such areas are also legislated to provide important spaces for human use, such as recreation and leisure activities. The Canada National Parks Act (2000), for example, explicitly states that, 'The national parks of Canada are hereby dedicated to the people of Canada for their benefit, education and enjoyment' (Section 4.1). Similarly, the Ontario Provincial Parks and Conservation Reserves Act (2006), states that one objective of provincial parks is 'To provide opportunities for ecologically sustainable outdoor recreation opportunities and encourage associated economic benefits' and that these areas 'are dedicated to the people of Ontario and visitors for their inspiration, education, health, recreational enjoyment and other benefits' (Government of Ontario 2006: Section 2.2).

Protected areas represent a significant tourist attraction for visitors to and from Canada. Canadian national parks alone attract more than 13 million visits per year, which translates into approximately 42 million visitor days of activity, and total visitation to provincial parks in the province of Ontario generally exceeds nine million annually with six of the last nine years exceeding ten million (Ontario Parks 2009). Climate change has potentially important implications for tourism opportunities in parks and other protected areas because visitor use is strongly correlated to climate. In order to understand how climate change could affect the future prospects of tourism and the relative competitiveness of tourism destinations (including protected areas), it is important to recognize that all of the major components of the global tourism system (tourists, source markets, transport systems, destinations) will be affected by four distinct impact pathways: (1) direct climatic changes (e.g., length and quality of tourism seasons, operating costs, increased infrastructure damage, and business interruptions); (2) indirect climate-induced environmental changes (e.g., water availability, biodiversity loss, reduced landscape aesthetic, increased natural hazards, coastal erosion and inundation, increasing incidence of vector-borne diseases); (3) indirect climate-induced social-economic changes (e.g., social unrest, political instability, decreased economic growth and discretionary wealth, altered environmental attitudes); and, (4) climate change mitigation policy (e.g., increase in transport costs, decreased accessibility to some destinations) (Scott and Lemieux 2011) (Figure 14.4).

In a protected areas context, climate influences the physical resources (e.g., water levels, snow cover, and wildlife species) that provide the foundation for outdoor recreation (e.g., boating, cross-country skiing, birdwatching), defines when specific activities can take place (e.g., beach use, swimming), and

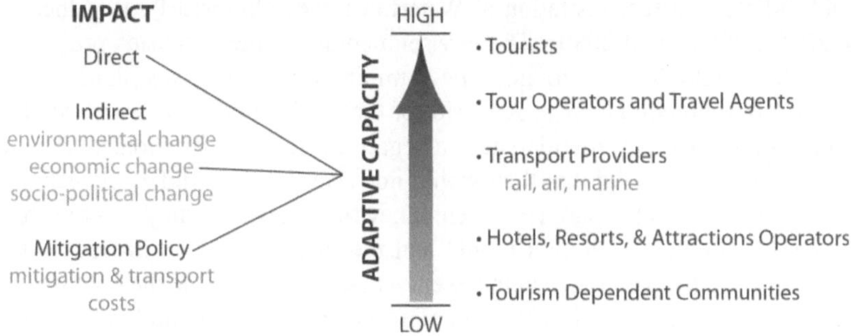


*Figure 14.4* Climate influences on the tourism sector and relative adaptive capacity of major tourism subsectors (adapted from Scott and Lemieux 2011 and Scott and Jones 2006)

influences the level of visitor satisfaction (Jones and Scott 2006a and 2006b). Two recent analyses indicate that Canada's protected areas could experience an increase in visitors under climate change due to a lengthened and improved warm-weather tourism season (Jones and Scott 2006a and 2006b). For example, Jones and Scott (2006a) found that overall visitation levels to Canadian parks could increase 6 to 8 percent in the 2020s, 9 to 29 percent in the 2050s and between 10 and 41 percent in the 2080s, with the largest increases in visitation occurring during the spring and fall months as climatic conditions conducive to warm-weather recreation activities persist for longer periods of time.

Some of the most significant increases in visitation are projected for national parks located in more northerly locations, such as Pukaskwa Provincial Park in Ontario (2020s: +12.2 to 22.6 percent; 2050s: +14.2 to 40.2 percent; 2080s: +16.4 to 58.8 percent) and Prince Albert National Park in Saskatchewan (2020s: +6.7 to 14.6 percent; 2050s: +10.4 to 35.7 percent; 2080s: +11.7 to 55.1 percent). Similar results are projected for Ontario's provincial parks (Jones and Scott 2006b) where visitation could potentially increase between 11 and 27 percent system-wide in the 2020s and between 15 and 56 percent in the 2050s. This shift is already taking place in Ontario Provincial Parks where there has been a substantial increase in visitation in the September and October shoulder season, largely attributed to warmer conditions for outdoor recreation (van Staalduinen pers. comm). In the 2080s, the number of people visiting Ontario's provincial parks is projected to increase between 19 and 82 percent (Jones and Scott 2006b).

Despite the findings of Jones and Scott (2006b) noted above, which revealed potential increases in visitation resulting from climate change, any changes in the natural characteristics of environments could also negatively influence tourism by reducing the perceived attractiveness of a region's parks. This is because climate indirectly affects nature-based tourism by impacting the physical resources that define the nature and quality of natural environments on which protected areas tourism depends (i.e., climate-induced biophysical change). A more recent study by Scott *et al.* (2007) explored how climate-induced environmental change could indirectly affect visitation at Waterton Lakes National Park (Alberta) for the 2020s, 2050s, and 2080s. The environmental change scenarios projected for the 2020s and 2050s were found to have minimal influence on visitation. However, the environmental change scenario for the 2080s (under the warmest climate change conditions) was found to have a negative effect on visitation, as 19 percent of respondents indicated that they would not visit the park and 37 percent stated that they would visit the park less often. A key finding of the study was the contrast between the climate-visitation model and the visitor survey with regard to the impact of climate change late in the twenty-first century. While the climate-visitation model projected the direct impact of a changed climate would increase visitation to the park (+11 percent to 60 percent for the 2080s), the survey found that the indirect impact of climate-induced environmental change in the park might reduce visitation, with 56 percent of respondents indicating that they would no longer come to the park or would visit less often if the environmental changes under the warmest 2080s climate change scenario were realized. A number of recent studies have similarly revealed that the willingness of tourists to revisit particular destinations is strongly linked to the state of particular environmental attributes (see, for example, Uyarra *et al.* 2005). Overall, Canada's national and provincial parks and other forms of protected areas are major resources for nature-based tourism and any changes in the length and quality of recreation seasons induced by climate change will have considerable implications for park visitation, revenue, and management requirements.

## Managing last chance tourism in protected areas

Similar to protected areas around the world, most protected areas in Canada were established with the intent of perpetually protecting highly valued individual species and their habitats. The primary goal of protected areas management and planning is to ensure that there is a clearly defined direction for the maintenance or restoration of ecological integrity and, in light of this primary goal, for guiding appropriate use of these areas. Changes in seasonal visitation rates will be an important issue for protected area managers in Canada because there are significant implications for user-fee collection, staffing needs, and environmental operations. While protected areas agencies would certainly benefit economically from increased visitors, an increase in visitors during the peak tourism season in some parks would also place extra stress on park assets (Scott and Lemieux 2007) while in others it would be welcome (i.e., in areas that are currently underutilized).

While managing last chance tourism resources and tourism visitation in protected areas will be necessary, many of the management challenges pertaining to climate change and last chance tourism are no different than those already being dealt with by protected areas agencies and organizations. It is likely that management strategies to mitigate the impacts of additional visitors will be needed in the short term to offset further ecological degradation and reduce potential conflicts among park users (Jones and Scott 2006a). As Eagles *et al.* (2002) emphasized, the principal question confronting park tourism planning is to determine what *degree* of impact is acceptable. The authors suggest that such decisions represent value judgments about the relative worth of the trade-offs involved. How much damage to the natural environment is worth the positive economic and quality of life gains from park tourism? To help answer such questions, Table 14.1 sets forth some key principles identified by Eagles *et al.* (2002) for protected areas visitor management which are transferrable to a last chance tourism context.

*Table 14.1* Principles of visitor management in protected areas (Adapted from: Eagles *et al.* 2002)

| *Principle* | *Description* |
|---|---|
| Appropriate management depends on objectives | • Objectives provide definitive statements of the outcomes of protected area management.<br>• They identify the appropriateness of management actions and indicate acceptable resource and social conditions.<br>• They allow evaluation of success of management actions.<br>• The specific objectives are likely to be more contentious than general value statements.<br>• The process of establishing objectives is essentially political; therefore, public participation is essential. |
| Diversity in resource and social conditions in protected areas is inevitable and may be desirable | • Impacts, use levels, and expectations of appropriate conditions tend to vary (e.g. impact of a camp-site in periphery vs. centre of the protected area).<br>• Environmental variables influence visitor use and level of impact (e.g. topography, vegetation, access).<br>• Managers can identify this diversity, then make decisions on its desirability, thereby separating technical decisions from judgmental ones.<br>• Using zoning explicitly to manage for diverse recreation opportunities is more likely to preserve important values. |
| Management is directed at influencing human-induced change | • Protected areas often protect natural processes as well as features, so management is generally oriented to managing human-induced change since it causes most disturbances.<br>• Human-induced change may lead to conditions considered to be undesirable.<br>• Some changes are desirable and may be the reason for the creation of the park. For example, many parks are created to provide recreation opportunities and local economic development.<br>• Management actions determine what actions are most effective in influencing amount, type and location of changes. |
| Impacts on resource and social conditions are inevitable consequences of human use | • Even small amounts of recreational use can lead to disproportionately large biophysical or social impacts, so any level or use leads to some impact.<br>• Many impacts are purposefully designed; for |

*Table 14.1* Continued

| *Principle* | *Description* |
|---|---|
| | example, providing a certain level of environmental education for park visitors.<br>• Managers must ask: 'How much impact is acceptable or desirable?'<br>• The process to determine the acceptability of impact is central to all visitor planning and management.<br>• Managers must utilize appropriate actions to create and manage this acceptable level of impact. |
| Impacts may be temporally or spatially discontinuous | • Impacts from visitor use or management activities may occur out of the protected area, or not be visible until later (e.g. prohibitions of use may displace that use to other areas; or poor water treatment may result in water pollution downstream).<br>• Planners need substantial knowledge of relationships between use and impacts to predict relationships at a variety of scales and over time. |
| Many variables influence the use/impact relationship | • Many variables other than level of use affect the use/impact relationship in protected areas (e.g. behavior of visitors, travel method, group size, season, and biophysical conditions).<br>• Education and information programs, as well as regulations aimed at restricting visitor behavior, may be necessary in addition to limits of use. |
| Many management problems are not dependent on numbers of users | • Management issues relating to the density of human use often have relatively simple technological solutions (e.g. parking, toilet facilities, water supply).<br>• But the relationship to use is not always linear. For example, the facilities designed for a few users may have very large impacts, but facilities designed for many more users have proportionally less additional impact.<br>• Similarly, social conditions (e.g. visitor satisfaction) are not always density-dependent. |
| Limiting use is only one of many management options | • A use-limit policy is only one of a number of potential management actions available, and is one of the most intrusive actions that protected area managers can employ.<br>• There are many issues involved in employing limits to use, such as choosing appropriate allocation or rationing techniques.<br>• Limiting use can have major political problems because of the necessary decision of who does not get access, and how access is allocated. |
| The decision-making process should separate technical decisions from value judgments | • Many protected area management decisions are technical (e.g. location of trail, design of visitor centre). But others reflect value judgments (e.g. decisions to limit use, and how, types of facilities, tourism opportunities provided).<br>• Decision processes should separate questions of 'existing conditions' from 'preferred conditions'. |

Managers will increasingly be challenged to reconcile the impacts of recreational activity and tourism with protection of the integrity of the natural resources upon which these experiences depend. Publicizing the vulnerability of a destination or vulnerable attributes of a destination can be viewed as a double-edged sword (Dawson *et al.* 2010). While last chance tourism can be used to attract more tourists seeking to experience something before it disappears, it can also be used to generate revenue and raise awareness about global environmental change and other sustainability issues in a specific and concrete setting.

For example, because many ecotourism and last chance tourism resources are located within the world's protected areas estate, user fees associated with last chance tourism could be used to invest in conservation and communications efforts. Revenue generated from visitation would also provide economic development opportunities to communities located in and near these areas. A virtuous management scheme based on demand, user fees, and conservation, could be used to support sustainable tourism within last chance tourism destinations with benefits that extend well beyond these areas (i.e., through conservation reinvestment and revenue sharing models).

It has also been suggested that last chance tourism experiences can motivate behavioral changes in tourists who visit these areas. The role of tourism in environmental education is well known, but how exactly last chance destinations are used for education efforts have only been explored in a nominal way. Protected areas in particular can be used to provide visible examples of how climate change can affect natural and cultural last chance tourism assets. Many protected areas jurisdictions around the world incorporate various environmental issues into visitor services programs, including interpretative talks and literature on natural and cultural history, and in websites containing educational materials on protected areas and contemporary management issues (e.g., Parks Canada's climate change website: see Parks Canada Agency 2011b) (Figure 14.5). McKie (2007) argued that educational programming as well as the 'pro-poor or pro-conservation' benefits of last chance tourism might outweigh the environmental impacts associated with visitation. However, the extent to which these activities and resources influence demand for last chance tourism or educate visitors on environmental issues, such as climate change, remains unknown. Because climate change presents significant and complex ecological, social, and economic challenges and opportunities to protected areas, agencies and organizations will be responsible for communicating these challenges and opportunities to an increasingly diverse and concerned public audience. Parks offer the potential to educate millions of visitors annually on climate change impacts and their implications for natural assets, and they provide the opportunity to enhance understanding of associated managerial responses.

Managers have at their disposal a wide array of strategies to manage the impacts of park tourism. A number of sophisticated frameworks have been developed in recent years to provide a structure for the management of protected area visitation and tourism (e.g., limits of acceptable change, visitor impact management, and visitor experience and resource protection) (see Eagles *et al.* 2002 for a useful

*Figure 14.5* Students viewing Sandhill Cranes (Grus canadensis) and reading interpretive signs at Great Sand Dunes National Park and Preserve, Colorado, USA. Protected areas provide unique opportunities for place-based communication strategies that can help visitors understand both climate change impacts and management responses. (Photo credit: C.J. Lemieux)

review). Their choice will be determined by the availability of resources (e.g., staffing, financing), any restrictions that legislation or agency policy places upon them, by the efficiency and appropriateness of the management strategy, and the resource implications (Eagles *et al.* 2002). Moreover, understanding the behavior of 'future tourists' under changing climatic and environmental conditions is important to understanding potential visitor trends at protected areas, especially with respect to last chance tourism opportunities (e.g., glacier viewing, birding, polar bear viewing, etc.). In an era of rapid climate change, managers will increasingly have to make decisions regarding the tourism opportunities provided by protected areas. These decisions will require important ethical analysis and careful examination of environmental and socio-economic trade-offs.

## Emerging ethical considerations for protected areas managers

The undesirable side-effects of climate change and tourism have led to the growing concern for the conservation and preservation of natural resources, human health and well-being, and the long-term economic viability of communities. As noted previously, the primary goal of protected areas management and planning in Canada is to ensure that there is a clearly defined direction for the maintenance or restoration of ecological integrity and, in light of this primary goal, for guiding appropriate use of these areas. In a changing climate, the utility of prescribed burning, species re-introduction, species translocation (i.e., assisted migration), and other management tools and techniques that have traditionally been employed (or have been recently considered in a climate change context) to achieve these park management objectives will need to be addressed. For example, there are a number of ethical issues associated with the translocation of species (also often referred to as 'assisted migration') (see Stewart *et al.* this volume). The translocation of species-at-risk to more suitable habitat, a strategy which is commonly proposed in the literature (e.g., McLachlan *et al.* 2007; Hugh-Guldberg *et al.* 2008), could be interpreted as inconsistent with maintaining ecological integrity if the species in question is not native to the destination region and could have a negative impact on other native species (Scott and Lemieux 2005). Agencies will need to decide whether to manage protected areas in a manner that forestalls undesirable impacts of ecosystem change, or using an approach that facilitates ecosystem change 'naturally', or through active adaptive measures such as species translocation or restoration to anticipated future conditions.

Moreover, protected area managers will need to begin thinking about the ethical implications and obligations to species – a 'hands-off' or 'non-interventionist' approach may not be a tenable approach if a highly valued species (such as a species-at-risk or charismatic mega-fauna) is unable to adapt naturally to climate change-related impacts. The rate and magnitude of climate change projected for the twenty-first century is likely to exceed many of the thresholds to which current species assemblages have become adapted, regardless of such management interventions. It is conceivable that not all species will be able to persist despite the best conservation efforts. As such, 'conservation triage' may emerge to be a critical tool in the prioritization and selection of which species to assist. Triage in a conservation context is the process of prioritizing the allocation of limited resources to maximize conservation returns, relative to the conservation goals, under a constrained budget (Botrill *et al.* 2008).

Protected areas, which are primarily managed for public good by various government authorities worldwide, will face considerable ethical management challenges with respect to last chance tourism. For example, the recent global trend is for governments to demand that protected areas earn an increasing proportion of their budget from tourism sources (Eagles and Hillel 2008). In Canada, Parks Canada and Ontario Parks, the oldest protected areas agencies in the country, were recently reorganized using business operating models that rely more heavily on revenue retention to support the management and operations of their respective protected area systems. Other key components of Ontario Parks' business model included a flattened organizational structure, increased flexibility in pricing policy, increased ability to enter into business partnerships with private corporations and pubic non-governmental organizations, and the ability to receive gifts.

While protected areas will remain a fundamental tool in the conservation of biodiversity in an era of global environmental change (Lemieux *et al.* 2011b), last chance tourism opportunities could lead to the emergence of conservation triage operations based on potential revenue generation and opportunities for business partnerships, thereby jeopardizing other conservation objectives and affecting the equitable treatment of park resources, such as vulnerable species that are less charismatic. This is especially troublesome when considering the recent estimates of the IPCC (2007) that approximately 20 to 30 percent of Earth's plant and animal species are likely to be at increased risk of extinction if increases in global average temperature exceed 1.5 to 2.5°C. The ethical dimensions of last chance tourism management (e.g., what species or other valued resources to save and the obligations to 'future generations' who will have to live with, and possibly bear the costs associated with, management decisions made today) clearly requires further deliberation in public policy on biodiversity conservation in Canada and elsewhere. The allocation of scarce resources to assist certain species (and not others) will have considerable ethical implications, and despite the burgeoning literature on climate change and biodiversity conservation over the past 20 years, debate and discussion over core questions such as 'Which species will be saved?' (i.e., use limited resources to avoid trying to save one species at the expense of several others?) and 'Is it acceptable to let a species go extinct in a national park?' have not yet occurred. In addition, the managers must be prepared to deal with new species introducing themselves naturally and also with major population shifts as new climate regimes provide better habitat conditions for some existing species. It is probable that the increases in mountain pine beetle populations in western Canadian forests and the Double-crested Cormorants in the Great Lakes maybe harbingers of coming population shifts due to climate change.

The goals of ecological restoration within protected areas, which are often aimed at re-establishing or maintaining native vegetation types (e.g., prescribed burning activities for prairie and savannah restoration), may similarly need re-assessment. The usefulness and viability of using historical ecosystem conditions as baseline reference points or targets must be evaluated in light of the fact that restoring these historic ecosystems may be more difficult, if not impossible, in the novel biophysical conditions of the future (e.g., Hobbs and Cramer 2008; Dunwiddie *et al.* 2009). In addition, such activities may prove to be an inefficient use of (limited) resources by protected areas agencies in relation to other priorities.

However, the question of whether it is ecologically, economically, and socially prudent to establish restoration programs with the objective of restoring areas to previous ecological conditions (which would possibly be attractive to last chance tourism pursuits) or, alternatively, develop objectives aimed at optimizing the health of ecosystems in a rapidly changing climate has not been deliberated in a structured and comprehensive manner. Moreover, how to provide ecological restoration within a broad cultural context (e.g., preventing 'new nature' from becoming 'non-places' devoid of cultural meaning and tourism attraction, or something so drastically different) is a social issue yet to be explored in Canada or elsewhere. More consideration and debate needs to be directed at defining the ecological, economic, and social (including ethical) implications of climate change for restoration practices within protected areas and then determining how restoration options relate to protected areas and tourism policy, planning, and management under new climate regimes and in light of new tourism trends such as last chance tourism.

Clearly, investigation into the complex trade-offs that exist between human well-being and protected areas goals, and between conservation and other economic, political, and social agendas across multiple scales will be required. While much attention has been given to bring ethics to tourism management, this has been primarily based on the development of 'codes of conduct'. In 1999, for example, The United Nations World Tourism Organization (UN-WTO) committed to a *Global Code of Ethics for Tourism* (UNWTO 1999). The code provides a framework designed to 'promote responsible, sustainable and universally accessible tourism' (UNWTO 1999: 2). Recognizing that all stakeholders in tourism – government, enterprises, business associations, workers, non-governmental organizations, host communities, the media, and the tourists themselves – have 'different albeit inter-dependent responsibilities in the individual and societal development of tourism', the code commits to WTO members to 'promote equitable, responsible and sustainable world tourism order' (UNWTO 1999: 3). More recently, the UN General Assembly adopted by consensus at its sixty-fifth session two separate resolutions emphasizing the role of tourism in sustainable development (Resolution 65/148: *Global Code of Ethics for Tourism*; and, Resolution 65/173: *Promotion of eco-tourism for poverty eradication and environment protection*) (UNGA 2011). The resolutions recognize the need to promote the development of sustainable tourism and increase the benefits from tourism for host communities, while maintaining their cultural and environmental integrity, and enhancing the protection of ecologically sensitive areas and natural heritages. Likewise, they recognize the need to promote the development of sustainable tourism and address the challenges of climate change and halt the loss of biodiversity.

## Conclusions

Owing to the size and diversity of Canadian ecosystems and the protected areas already established to represent them, climate change may present itself as a spectral disorder across this unique Canadian mosaic. Although climate change may drastically alter the character and ecological composition of some protected areas, these areas will continue to protect important segments of Canada's ecosystems and biodiversity for scientific study and heritage appreciation.

While the volume of climate change adaptation research has increased significantly in the last five years, response by the protected areas and tourism disciplines is lagging. It may be because of the complexity and uncertainty of managing protected areas in a rapidly changing climate (Scott and Lemieux 2005). Unlike other managed resource systems (e.g., water, agriculture, fisheries), there are no past exposures or climate change analogues to learn from at the systems planning or protected area management levels. Furthermore, the management objectives for protected areas have very long time horizons (e.g., twenty-second century and beyond), and fewer adaptation options exist for protected areas than for lands and waters that are actively and extensively manipulated or managed for a variety of other human uses.

Despite these uncertainties, managers and planners need to recognize that climate change will simultaneously present challenges and opportunities for tourism within Canada's protected areas. Resources within protected areas that are currently attractive to last chance tourism pursuits will increase and then disappear; just the same as new opportunities will emerge and then change. Currently, there is little understanding of how climate change will stimulate a diversification of tourism products and destinations within and outside of protected areas. Clearly, there is a need to better understand how climate change could affect last chance tourism attractions, visitation patterns, and the overall quality of visitor experiences within protected areas. Likewise, protected areas managers will need to determine how to manage for climate change-induced visitor behavioral responses and shifting (and emerging) last chance tourism destinations and preferences (e.g., how to adapt infrastructure, accommodate losses and emergence of destination preferences, and ensure visitor safety). Additional analysis that examines how climate change may interact with other factors (e.g., population growth, ageing society, travel costs) to impact tourism in protected areas will also be fundamental. Indeed, the issues related to the strategic response of protected area agencies to the challenge of climate change are very complex and deserve further elaboration than that provided here.

In summary, across Canada protected areas have assumed an important role in efforts to conserve biodiversity and to enhance ecological sustainability within and beyond formally designated sites. Climate change impacts are already discernable within the global protected areas estate and the role of protected areas in conserving biodiversity will become even more important over the twenty-first century and beyond. Tourism in and around protected areas must be designed as a vehicle for conservation: building support, raising awareness of the many important values of protected areas (including ecological, cultural, spiritual, aesthetic, recreational, and economic) and generating income for work aimed at conservation and enhancing biodiversity, ecological integrity, and cultural heritage. Research is needed in this area and it should be linked to corporate efforts of developing, managing, and monitoring sustainable business practices from an ethical perspective in the marketplace, society, and natural world.

Decisions related to climate change and tourism in protected areas will, in the end, come down to value judgments. Tourism managers and operators, and other landowners, be they public or private, will need to work closely with those responsible for natural heritage areas, regional planning, and environmental quality to ensure that last chance tourism activities in parks and other forms of protected areas are managed effectively. Their collective cooperation should ensure that the twin objectives of protection and development are suitably balanced in line with sustainability goals outlined in legislation and policy while also working towards improving the capacity of protected areas agencies and organizations worldwide to understand (via research and monitoring) and adapt to climate change-related impacts. A constant influx of new information about climate change, ecosystem function, and how ecosystem change will affect visitor preferences will help practitioners and other stakeholders make rational decisions that balance protection and development or economic mandates. Accordingly, an important aspect of any such enquiry is the need to document and report case studies on various adaptive approaches and techniques initiated to deal with climate change so that knowledge and experiences are shared among practitioners.

## References

Araújo, M.B., Cabeza, M., Thuiller, W., Hannah, L., and Williams, P. (2004) 'Would climate change drive species out of reserves? An assessment of existing reserve-selection methods', *Global Change Biology* 10(9): 1618–1626.

Araújo, M.B. (2009a) 'Protected areas and climate change in Europe' (Draft Version), *Convention on the Conservation of European Wildlife and Natural Habitats*, Bern Switzerland.

Araújo, M.B. (2009b) 'Climate change and spatial conservation planning', in H.P.A. Moilanen and K. Wilson (eds.) *Spatial Conservation Prioritization: Quantitative Methods and Computational Tools*, Oxford: Oxford University Press.

Baron, J.S., Gunderson, L., Allen, C.D., Fleishman, E., McKenzie, D., Meyerson, L.A., Oropeza, J., and Stephenson, J. (2009) 'Options for national parks and reserves for adapting to climate change', *Environmental Management* 44(6): 1033–1042.

Bartlein, P.J. (1997) 'Past environmental changes: characteristic features of Quaternary climate variations', in B. Huntley, W. Cramer, A.V. Morgan, H.C. Prentice, and J.R.M. Allen (eds.) *Past and Future Rapid Environmental Changes: The Spatial and Evolutionary Responses of Terrestrial Biota*, NATO ASI Series, vol. 147.

Blais, J., Schindler, D., Muir, D., Kimpe, L., Donald, D., and Rosenburg, B. (1998) 'Accumulation of persistent organochlorine compounds in mountains of western Canada', *Nature* 395: 585–588.

Bomhard, B., Richardson, D.M., Donaldson, J.S., Hughes, G.O., Midgley, G.F., Raimondo, D.C., Rebelo, A.G., Rouget, M. and Thuiller W. (2005) 'Potential impacts of future land use and climate change on the Red List status of the Proteaceae in the Cape Floristic Region, South Africa', *Global Change Biology* 11: 1452–1468.

Bottrill, M., Joseph, L., Carwardine, J., Bode, M., Cook, C., Game, E., *et al.* (2008) 'Is conservation triage just smart decision-making?' *Trends Ecol. Evol.* 23: 649–654.

Burns, C.E., Johnston, K.M. and Schmitz, O.J. (2003) 'Global climate change and mammalian species diversity in U.S. national parks', Proceedings of the National Academy of Sciences (PNAS) 100(20): 11474–11477.

Bushell, R. and Eagles, P.F.J. (2007) *Tourism and Protected Areas: Benefits Beyond Boundaries,* Cambridge: IUCN and CABI.

Canadian Council on Ecological Areas (2011) *Conservation Areas and Reporting Tracking System (CARTS).* Gatineau, Québec, Canada: Canadian Council on Ecological Areas (CCEA) and Environment Canada. Online: www.ccea.org/en_carts.html (accessed 3 April 2011).

Coulston, J.W. and Riitters. K.H. (2005) 'Preserving biodiversity under current and future climates: A case study', *Global Ecology and Biogeography Letters* 14: 31–38.

Dawson, J., Stewart, E.J., Lemelin, H. and Scott, D. (2010) 'The carbon cost of polar bear viewing in Churchill, Canada', *Journal of Sustainable Tourism* 18: 319–336.

Dunlop, M. and Brown, P.R. (2008) *Implications of Climate Change for Australia's National Reserve System: a Preliminary Assessment,* CSIRO Sustainable Ecosystems, report to the Department of Climate Change and the Department of the Environment, Water, Heritage and the Arts, Canberra, ACT.

Dunwiddie, P.W., Hall, S.A., Ingraham, M.W., Bakker, J.D., Nelson, K.S., Fuller, R. and

Gray, E. (2009) 'Rethinking conservation practice in light of climate change', *Ecological Restoration* 27(3): 320–329.
Eagles, P.F.J., McCool, S.F. and Haynes, C.D.A. (2002) *Sustainable Tourism in Protected Areas: Guidelines for Planning and Management.* Cambridge: IUCN Gland, Switzerland.
Eagles, P. and Hillel. O. (2008) 'Improving protected area finance through tourism', in *Protected Areas in Today's World: their values and benefits for the welfare of the planet,* Secretariat of the Convention on Biological Diversity, Technical Series no. 36, p 77–86. Online: www.cbd.int/doc/publications/cbd-ts-39-en.pdf (accessed 4 April 2011).
ENPAA (English National Park Authorities Association) (2009) *Climate Change Mitigation and Adaptation in National Parks.* London: ENPAA.
Government of Canada (2000) Canada National Parks Act (S.C. 2000, c. 32).
Government of Canada (2002) Species at Risk Act (S.C. 2002, c. 29).
Government of Ontario (2006) Provincial Parks and Conservation Reserves Act (S.C. 2006, c. 12).
Gray, P.A. (2005) 'Impacts of climate change on diversity in forested ecosystems: Some examples', *Forestry Chronicle* 81(5): 655–661.
Griffith, B., Scott, J.M. Adamcik, R., Ashe, D., Czech, B., Fischman, R., Gonzalez, P., Lawler, J., McGuire, A.D. and Pidgorna. A. (2009) 'Climate change adaptation for the U.S. National Wildlife Refuge system', *Environmental Management* 44(6): 1043–1052.
Hall, M.P. and Fagre, D.B. (2003) 'Modeled climate-induced glacier change in Glacier National Park 1850–2100', *Bioscience* 53(2): 131–140.
Hannah, L. and Salm. R. (2005) 'Protected areas management in a changing climate', in T.E. Lovejoy and L. Hannah (eds.) *Climate Change and Biodiversity.* London: Yale University Press.
Hobbs, R.J. and Cramer. V.A. (2008) 'Restoration ecology: Interventionist approaches for restoring and maintaining ecosystem function in the face of rapid environmental change', *Annual Review of Environment and Resources* 33: 39–61.
Hugh-Guldberg, O., Hughes, L., McIntyre, S., Lindenmayer, D.B., Parmesan, C., Possingham, H.P., and Thomas. C.D. (2008) 'Assisted colonization and rapid climate change', *Science* 321: 345–346.
Huntley, B. (2007) *Climatic Change and the Conservation of European Biodiversity: Towards the Development of Adaptation Strategies* – discussion paper. Durham: Durham University, Institute of Ecosystem Science, School of Biological and Biomedical Sciences.
IPCC (Intergovernmental Panel on Climate Change) (2007) *Climate Change 2007: The Physical Science Basis,* Working Group I Contribution to the Fourth Assessment Report of the Intergovernmental Panel on Climate Change (IPCC) Fourth Assessment Report, Cambridge: Cambridge University Press.
IUCN and UNEP-WCMC (World Conservation Union and the United Nations Environment Programme World Conservation Monitoring Centre) (2011) *World Database on Protected Areas*, (WDPA). Cambridge: IUCN, UNEP-WCMC. Online: www.wdpa.org/ (accessed 2 April 2011).
Jones, B. and Scott. D.J. (2006a) 'Climate change, seasonality and visitation to Canada's national parks', *Journal of Parks and Recreation Administration* 24(2): 42–62.
Jones, B. and Scott, D.J. (2006b) 'Implications of climate change for visitation to Ontario's provincial parks', *Leisure* 30(1): 233–261.
Joyce, L., Blate, G.M., McNulty, S.G., Millar, C.I., Moser, S., Neilson, R.P. and Peterson,

D.L. (2009) 'Managing for multiple resources under climate change: National forests', *Environmental Management* 44(6): 1022–1031.

Lawler, J.J., Shafer, S.L., White, D., Kareiva, P., Maurer, E.P., Blaustein, A.R. and Bartlein,. P.J. (2009) 'Projected climate-induced faunal change in the Western Hemisphere', *Ecology* 90(3): 588–597.

Lemelin, H., Dawson, J., Stewart, E.J., Maher, P. and Lück. M. (2010) 'Last-chance tourism: the boom, doom, and gloom of visiting vanishing destinations', *Current Issues in Tourism* 13(5): 477–493.

Lemieux, C.J. and Scott. D.J. (2005) 'Climate change, biodiversity conservation and protected area planning in Canada', *Canadian Geographer* 49(4): 384–399.

Lemieux, C.J., Beechey, T.J., Scott, D.J. and Gray, P.A. (2010) *Protected Areas and Climate Change in Canada: Challenges and Opportunities for Adaptation*, Technical Report No. 19. Ottawa, Canada: Canadian Council on Ecological Areas (CCEA).

Lemieux, C.J., Beechey, T.J., Scott, D.J. and Gray, P.A. (2011a) 'The state of climate change adaptation in Canada's protected areas sector', *Canadian Geographer* doi: 10.1111/j.1541–0064.2010.00336.x.

Lemieux, C.J., Beechey, T.J. and Gray, P.A. (2011b) 'Prospects for Canada's protected areas in an era of climate change', *Land Use Policy*. doi: 10.1016/j.landusepol.2011.03.008.

Luckman, B. and Kavanagh, T. (2000) 'Impact of climate fluctuations on mountain environments in the Canadian Rockies', *Ambio* 29(7): 371–379.

Malcolm, J.R., Liu, C., Miller L.B., Allnutt, T. and Hansen. L. (2002) *Habitats at Risk: Global Warming and Species Loss in Globally Significant Terrestrial Ecosystems,* Gland Switzerland: Worldwide Wildlife Fund (WWF).

Malcolm, J.R., Canran, L., Neilson, R.P., Hansen, L. and Hannah. L. (2006) 'Global warming and extinctions of endemic species from biodiversity hotspots', *Conservation Biology* 20(2): 538–548.

Mawdsley, J.R., O'Malley, R. and Ojima. D.S. (2009) 'A review of climate-change adaptation strategies for wildlife management and biodiversity conservation', *Conservation Biology* 23(5): 1080–1089.

McKenney, D., Pedlar, J.H., Lawrence, K., Campbell, K. and Hutchinson. M.F. (2007a) 'Potential impacts of climate change on the distribution of North American trees', *BioScience* 57(11): 939–948.

McKenney, D., Pedlar, J.H., Lawrence, K., Campbell, K. and Hutchinson, M.F. (2007b) 'Beyond traditional hardiness zones: using climate envelopes to map plant range limits', *BioScience* 57(11): 929–937.

McKie, R. (2007) '10 wonders of the vanishing world', *The Observer*. Online: www.guardian.co.uk/travel/2007/jan/28/wildlifeholidays.ethicalholidays.green (accessed 28 January 2011).

McLachlan, J., Hellmann, J. and Schwartz. M. (2007) 'A framework for debate of assisted migration in an era of climate change', *Conservation Biology* 21: 297–302.

McLaughlin, J.F., Hellmann, J.J., Boggs, C.L. and Ehrlich, P.R. (2002) 'Climate change hastens population extinctions', *PNAS* 99(9): 6070–6074.

NPS (National Park Service) (2010) *National Park Service Climate Change Response Strategy.* Fort Collins, CO: National Park Service Climate Change Response Program.

Ontario Parks (2009) *Park Statistics/Statistiques Parcs 2009*. Peterborough, Ontario: Ontario Ministry of Natural Resources (OMNR). Online: www.OntarioParks.com/statistics/2009_park_statistics.pdf (accessed 2 April 2011).

Parks Canada Agency (2009) *Visitor Attendance Records at National Parks.* Online: www.pc.gc.ca/eng/docs/pc/attend/index.aspx (accessed 2 April 2011).

Parks Canada Agency (2011a) *Impacts On Western Cordillera Parks*. Online: www.pc.gc.ca/docs/v-g/ie-ei/cc/repercussions-impacts2.aspx (accessed 2 April 2011).

Parks Canada Agency (2011b) *The Climate is Changing in Our National Parks*. Online: www.pc.gc.ca/docs/v-g/ie-ei/cc/actions.aspx (accessed 2 April 2011).

Parmesan, C. and Yohe, G. (2003) 'A globally coherent fingerprint of climate change impacts across natural systems', *Nature* 41: 37–42.

Parmesan, C. (2006) 'Ecological and evolutionary responses to recent climate change', *Annual Review of Ecology, Evolution, and Systematics* 37: 637–69.

Pounds, J.A., Fogden, M.P.L. and Campbell. J.H. (1999) 'Biological responses to climate change on a tropical mountain', *Nature* 398: 611–615.

Root, T.L., Price J.T., Hall, K.R., Schneider, S.H., Resenzweig, C. and Pounds. J.A. (2003) 'Fingerprints of global warming on wild animals and plants', *Nature* 421: 57–60.

Rutherford, M.C., Powrie, L.W. and Schulze, R.E. (1999) 'Climate change in conservation areas of South Africa and its potential impact on floristic composition: a first assessment', *Diversity and Distributions* 5: 253–262.

Schwartz, M.W., Iverson, L.R., Prasad, A.M., Matthews, S.N. and O'Connor. R.J. (2006) 'Predicting extinctions as a result of climate change', *Ecology* 87: 1611–1615.

Scott, D.J. and Suffling. R. (2000) *Climate Change and Canada's National Parks*. Toronto: Environment Canada.

Scott, D.J., Malcolm, J.R. and Lemieux, C.J. (2002) 'Climate change and modeled biome representation in Canada's national park system: Implications for system planning and park mandates', *Global Ecology and Biogeography* 11: 475–484.

Scott, D.J. and Lemieux, C.J. (2005) 'Climate change and protected area policy and planning in Canada', *Forestry Chronicle* 81(5): 696–702.

Scott, D.J. and Lemieux, C.J. (2007) 'Climate change and protected areas policy, planning and management in Canada's boreal forest', *Forestry Chronicle* 83(3): 347–357.

Scott, D. and Jones, B. (2006) *Climate Change and Seasonality in Canadian Outdoor Recreation and Tourism – Executive Summary*, report prepared for the Government of Canada Climate Change Action Fund. Waterloo, Ontario: University of Waterloo.

Scott, D.J., Jones, B. and Konopek. J. (2007) 'Implications of climate and environmental change for nature-based tourism in the Canadian Rocky Mountains: a case study of Waterton Lakes National Park', *Tourism Management* 28(2): 570–579.

Scott, D.J. and Lemieux, C.J. (2011) 'The vulnerability of tourism to climate change', in A. Holden, A. and D. Fennell (eds.) *Handbook for Tourism and Environment*. London: Routledge.

State of the Canadian Cryosphere (2011) *Past Variability in Canadian Glaciers*, Waterloo, Ontario: Department of Geography, University of Waterloo. Online: www.socc.uwwaterloo.ca (accessed 2 April 2011).

Téllez-Valdés, O. and Dávila-Aranda, P. (2003) 'Protected areas and climate change: a case study of the cacti in the Tehuacén-Cuicatlán Biosphere Reserve, México', *Conservation Biology* 17(3): 846–853.

Thuiller, W., Broennimann, O., Hughes, G.O., Alkemade, J.R.M., Midgley, G.F. and Corsi, F. (2006a) 'Vulnerability of African mammals to anthropogenic climate change under conservative land transformation assumptions', *Global Change Biology* 12: 424–440.

Thuiller, W., Lavorel, S., Sykes, M.T. and Araújo. M.B. (2006b) 'Using niche-based modelling to assess the impact of climate change on tree functional diversity in Europe', *Diversity and Distributions* 12: 49–60.

Thomas, C.D., Cameron, A., Green, R.E., Bakkenes, M., Beaumont, L.J., Collingham, Y.C., Erasmus, B.F.N., de Siqueira, M.F., Grainger, A., Hannah, L., Hughes, L., Huntley,

B., van Jaarsveld, A.S., Midgley, G.F., Miles, L., Ortega-Huerta, M.A., Peterson, A.T., Phillips, O.L. and Williams. S.E. (2004) 'Extinction risk from climate change', *Nature* 427: 145–148.

Thomas, C.D., Franco, A.M.A. and Hill. J.K. (2006) 'Range retractions and extinction in the face of climate warming', *Trends in Ecological and Evolution* 21(8): 415–416.

U.K. Climate Impacts Program (2001) *Climate Change and Nature Conservation in Britain and Ireland. MONARCH – Modelling Natural Resource Responses to Climate Change.* Summary Report, Oxford: UKCIP.

UN (United Nations) (2005) *Millennium Ecosystem Assessment: Ecosystems and Human Well-Being-Biodiversity Synthesis.* Washington, DC: World Resources Institute.

UNGA (United Nations General Assembly) (2011) *Resolutions of the UNGA 65th Session.* Online: www.un.org/en/ga/65/resolutions.shtml (accessed 2 April 2011).

UNWTO (United Nations World Tourism Organization) (1999) *Global Code of Ethics for Tourism, adopted by resolution A/RES/406(XIII) at the thirteenth WTO General Assembly*, Santiago, Chile, 27 September–1 October 1999. Online: www.unwto.org/ethics/index.php (accessed 2 April 2011).

Uyarra, M.C., Côté, I.M., Gill, J.A., Tinch, R.R.T., Viner, D. and Watkinson, A.R. (2005) 'Island-specific preferences of tourists for environmental features: Implications of climate change for tourism-dependent states', *Environmental Conservation* 1(32): 11–19.

van Staalduinen, B. (2011) Pers. Comm. Manager, Park Operations and Development Section. Peterborough, Ontario: Ontario Parks.

Welch, D. (2005) 'What should protected areas managers do in the face of climate change?', *George Wright Forum* 22(1): 75–93.

West, J.M., Julius, S.H., Kareiva, P., Enquist, C., Lawler, J.J., Peterson, B., Johnson, A.E. and Shaw. M.R. (2009) 'U.S. natural resources and climate change: Concepts and approaches for management adaptation', *Environmental Management* 44(6): 1001–1021.

Williams, S.E., Bolitho, E.E. and Fox. S. (2003) 'Climate change in Australian tropical rainforests: an impending environmental catastrophe', *Proceeding of the Royal Society of London,* 270: 1887–1892.

WCPA (World Commission on Protected Areas) (1993) *Parks for Life: Report of the IVth World Congress on National Parks and Protected Areas*, Gland, Switzerland: IUCN.

WCPA (World Commission on Protected Areas) (2003) *Benefits Beyond Boundaries: Proceedings of the Vth IUCN World Parks Congress,* Gland, Switzerland: IUCN.

# 15 Last chance tourism

## Conclusion

*Jackie Dawson, Emma Stewart, and Raynald Harvey Lemelin*

Relevance remains even centuries after the phrase 'the only constant in life is change' was coined by Greek philosopher Heraclitus of Ephesus, particularly as we move through the twenty-first century – an era being defined by continuous and rapid global change. To some, change represents a sense of newness, excitement and novel beginnings, while to others its mere suggestion embodies terror, fear, destruction, loss of control and death. It is no surprise then that the notion of last chance tourism, a form of tourism that capitalizes on that which is vulnerable and changing, evokes a polarity of opinions, which in this text have been expressed through discussions focused on (1) understanding, (2) marketing, and (3) managing last chance tourism.

### Understanding last chance tourism: disappearing landscapes/seascapes and vanishing fauna

As noted by Smith (this volume) and also by Lemelin and Baikie (this volume), the concept of last chance tourism is not necessarily new and has to some extent been evident throughout history. There are numerous examples of 'last chances to see'. For instance: thousands of tourists travelled to Germany to witness the 'last' of the Berlin wall before it was felled between 1989 and 1990; 1,700 passengers paid hefty fairs to be part of the well-publicized 'final voyage' of the QE2 (Queen Elizabeth II) ocean liner that travelled from South Hampton, England to New York, USA during the summer of 2008; in the same year thousands of fans packed Yankee stadium to experience the final game played in the historic park before it was demolished. Olsen, Koster and Youroukos (this volume) also note that last chance tourism might additionally include existing and seemingly durable locations such as the Forbidden City in China, the Taj Mahal in India, and the Pyramids of Giza in Egypt, because over-visitation by tourists, urbanization, and pollution have degraded the cultural and built fabric of these sites so much so that tourists perceive the authenticity of the attraction to be 'lost'. This perspective was also expressed by Johnston, Viken and Dawson (this volume) and Lemelin and Baikie (this volume) who discuss the impacts of globalization and modernization on Indigenous peoples. In some cases, this has led tourists to believe that their chance to see genuine Indigenous culture is already gone. Lamers, Eijgelaar and Amelung (this volume) make note of a number of additional circumstances that contribute to the creation of last chance tourism including infrastructure development, biological invasions, economic recessions, and policy changes. Each of these circumstances can cause a destination or tourism attraction to 'change' and can thus lead to a perception of vulnerability which can serve to either increase or decrease its 'value'.

A common theme emerging throughout this book is the strong notion that last chance tourism exists because of this perception of rarity and/or vulnerability (see the introduction for a discussion on the anthropogenic allee effect). Yet, last chance tourism not only requires a perception of vulnerability, there also needs to be a motivation amongst individual tourists to physically experience what is vulnerable. Steiger, Dawson and Stotter (this volume) reiterate this notion suggesting that the future implications of last chance tourism will depend heavily on the behavioural responses of individuals to change.

Dawson *et al.* (2010) envision the development of an inverse relationship between the supply of a vulnerable last chance tourism resource and the demand for last chance tourism opportunities (Figure 15.1). For example, as supply-side vulnerability increases, demand is also likely to increase, but subsequently is expected to decrease as the quality or reliability of the tourism asset (supply) diminishes even further or 'disappears'. In this scenario there is likely to be a temporal lag-effect whereby demand diminishes later than the observed decrease in supply, as may be observed with regard to wildlife viewing, particularly polar bear viewing (see Dawson *et al.* 2010; Stewart *et al.* this volume). Dawson *et al.* (in press) propose that the destination life cycle of last chance tourism is

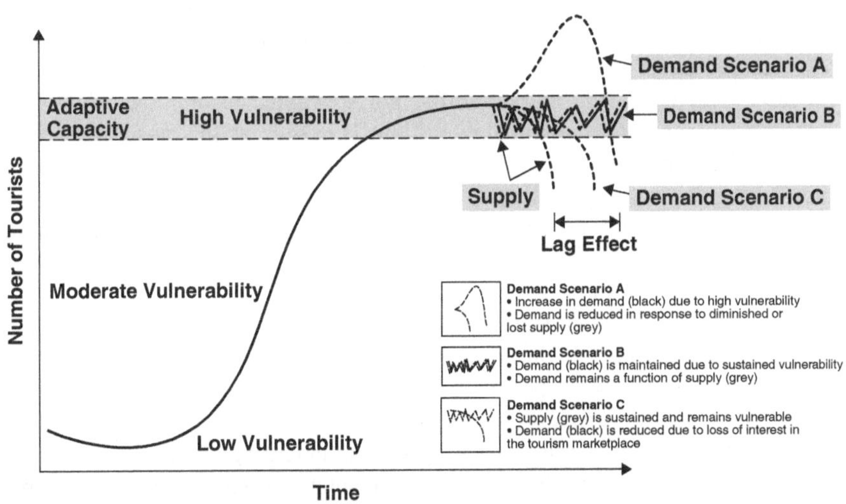


*Figure 15.1* Last chance tourism destination lifecycle scenarios (Dawson *et al.* in press)

likely to take on one of several different forms in response to changing vulnerability (i.e. supply) and demand and propose two additional scenarios labelling the original projection outlined above as scenario 'A'. In scenario 'B' the tourism asset is able to sustain its supply (or quality) and remain in a perpetual state of fluctuating vulnerability but never actually disappears or transforms into a new system (e.g., due to stagnation resulting from a geographic shift in the distribution of a vulnerable species or due to climate change mitigation and adaptation efforts that help deter extinction or maintain tourism access). In this situation it is possible that demand will be sustained and remain relative to supply thus fluctuating up and down as vulnerability increases and decreases, as alluded to by Coghlan (this volume) with regard to the Great Barrier Reef. Alternatively, demand scenario 'C' suggests that despite the perpetual vulnerability of supply, demand may at some point drop off due to a potential loss of interest within the tourist marketplace (e.g., due to inaccessibility or loss of novelty/value among tourists) even despite its continued vulnerability, as may be the case for the Kakapo reported by Shelton (this volume) (i.e. due to inaccessibility) (again see Figure 15.1).

Another theme emerging from the chapters in this volume is the idea that disappearing tourism resources within a particular region can actually translate into tourism opportunities in other regions, and in some cases can actually catalyse the development of new tourism enterprises entirely. This finding is not unlike recent discourse taking place within the more specific climate change and tourism literature whereby many authors suggest that the impact of climate change on tourism resources (i.e. supply) combined with the behavioural response of tourists themselves (i.e. demand) will influence the emergence of both 'winners' and 'losers' within a region or within a particular tourism sector (see Aall and Høyer 2005; Hamilton *et al.* 2005; Scott *et al.* 2008).

Both Hvenegaard and Steiger *et al.* (this volume) point out that at a regional scale the disappearance, or extreme transformation of vulnerable tourism resources at one destination can generate opportunities at other destinations because of demand-side market-shifts. For example, a common behavioural response to changed recreational conditions (i.e. such as the closure of a local ski area, the disappearance of glaciers or declining birding conditions) involves substituting the activity that is no longer available for an all-together different activity. This has for over 20 years been known as 'activity substitution' as defined within the theory of substitutability developed by Iso-Ahola (1986). Other substitution responses involve changing the amount of time one participates in the activity or substituting the place where the activity usually takes place (i.e. destination substitution) (Steiger *et al.* this volume; see also Stewart *et al.* this volume). These behavioural substitution options represent possible demand-side reactions to last chance tourism.

Johnston *et al.* (this volume) identify several emerging 'first chance tourism' enterprises that have directly developed in response to changing environmental conditions and examples of changing tourism landscapes. For example, they point to new opportunities that have emerged as the ice-edge in Greenland has receded, including the establishment of a significant boat-based tourism industry as well as the notable possibility of paddling-based voyages to the North Pole instead of the more traditional ski-based trip. Indeed there is an emerging dialectic between the notions of 'last chance tourism' and 'first chance tourism'. The tension is perhaps best exemplified by the recent and rapid emergence of a now flourishing Arctic cruise ship industry which has developed in response to changing climate conditions where the reduction of sea-ice has made previously inaccessible areas more accessible even to moderately ice-strengthened vessels (Stewart *et al.* 2011). The development of the Arctic cruise ship industry (i.e. supply of cruise offerings) is an example of first chance tourism (i.e. emerging opportunities associated with changing sea ice conditions), but the demand for Arctic cruise opportunities is in large part reflective of a desire amongst tourists to view changing Arctic landscapes, features and wildlife that are regularly identified as vulnerable and in danger of disappearing (see Figure 15.2). Thus the supply of Arctic cruises can be seen as first chance tourism whereas the demand-side response is an example of last chance tourism.

## Marketing and the role of media in last chance tourism

Over the past several years a variety of media sources have been increasingly publishing articles outlining the vulnerability of the world's best-loved natural phenomena and encouraging tourists to visit various natural sites 'before they disappear'. The editor of a travel agents' magazine use the phrase 'environmental tourists of doom' to describe what authors of this volume have been calling 'last chance tourists'. Various terms have been used interchangeably within these media reports to describe last chance tourism including: catastrophe tourism (Todras-Whitehill, 2007), climate tourism and extinction tourism (Leahy, 2008), climate change voyeurism (Intelligent Travel, 2007), climate sightseeing (JWT, 2007), doom tourism (Salkin, 2007), doomsday tourism and tourism of doom (Shipman, 2007), endangered destination tourism (Ruiz, 2008), and global warming tourism (McCarthy, 2007).

The recent swell of media attention promoting last chance tourism in relation to vulnerable landscapes combined with heightened anxiety over the potential impacts of climate change has challenged the appropriateness of promoting or explicitly marketing last chance tourism – a form of travel that may be blamed for contributing to the further destruction of already vulnerable destinations. Eijgelaar *et al.* (2010) and Dawson *et al.* (2010) make specific note of the disproportionate amount of greenhouse gas emissions last chance tourists are responsible for during long-haul travel to identified last chance destinations such as Antarctica and Churchill in Manitoba, Canada (claimed to be the 'polar bear viewing capital of the world'). The possibility of purchasing the chance to be the last one to see or do something in a place that may not be there in the near future can be quite alluring for some individuals. Thus it is not surprising that several authors in this book have highlighted concern that some tour operators may be taking advantage of the surge in interest in the impact of global warming by providing 'last chance tours'

*Figure 15.2* Last chances to see melting glaciers in the Canadian Arctic from the Kapitan Khlebnikov (Photo credit: Emma J. Stewart)

to vulnerable sites. However, in a study of the online marketing of last chance destinations, Frew (2008, this volume) found that although particular sectors of the media have been using aspects of global warming and disappearing natural attractions as interesting and exciting angles for their stories (see list above), in reality few tour operators are explicitly using the last chance concept to promote their products. In this regard, it seems the existence of last chance tourism is based more on media publicity than it is on overt marketing conducted by tourism operators. This perspective is particularly highlighted by Olsen *et al.* (this volume) and Johnston *et al.* (this volume), who through a series of interviews with tourism operators note a strong reluctance towards the use of the last chance tourism concept within any aspect of their marketing strategies. One respondent in the study conducted by Olsen *et al.* (this volume) aptly stated; 'I think people should visit this beautiful area to participate in its preservation, not attend its funeral'.

## Ethics, culture, and management of last chance tourism

Despite legitimate concern amongst tourism operators regarding the consequences of a last chance tourism designation, publicizing the idea (as has been done extensively through popular media sources) can in some instances influence an influx of conservation funds and a 'call for action'. Lemieux and Eagles (this volume) note that last chance tourism resources are often located within the world's protected area estates (or are protected under other legislation, e.g. species-at-risk Act) and user fees associated with last chance tourism could be used to invest in conservation and education efforts. Revenue generated from visitation could also provide economic development opportunities or transition funding for communities located in or near these areas. Coghlan (this volume) also note a correlation between the perceived vulnerability of the Great Barrier Reef and the amount of conservation funds raised for the region.

As Shelton's chapter demonstrates, the perceived rarity of the kakapo stimulated national conservation efforts including ownership restrictions, awareness campaigns appealing to consumer ethics. In other places, the provision of sustainable alternatives through captive breeding and release programmes such as the Bali starling project in the Teluk Telor, Indonesia (see Newsome and Rodger this volume). In other situations, visitor impacts can be regulated through codes of conducts, marked trails, and viewing platforms such as those used at the Sepilok Orang-utan Rehabilitation Centre in Southeast Asia (Newsome and Rodger, this volume), educational initiatives and interpretation signs such as those used in protected areas (Lemieux and Eagles, this volume), or educational programmes offered to polar bear viewing tourists in Churchill, Manitoba (Lemelin 2006). Some last chance tourism destinations may require more dramatic actions including trade restrictions and fines enforced through local laws and/or by international treaties such as the Convention on International Trade in Endangered Species of Wild Fauna and Flora (CITES) or through the development and management of specially designated protected areas (see Lemieux and Eagles this volume). In other situations management approaches such as guided patrols and user-fees may be combined. In the case of the International Gorilla Conservation Programme, the income from tourism for the gorillas has been the most important factor in contributing to the support of parks and habitat conservation (Kirby 2002).

Information strategies aimed at reminding the visitors of the vulnerability of these sites, may reduce the impact on the species that are the targets from these activities and may, as Frew, Lamers *et al.* and Lemieux and Eagles (this volume) each suggest, encourage visitors to be become goodwill ambassadors. This approach is meant to raise the awareness of travellers encouraging them 'to take steps towards more responsible and sustainable everyday behaviours' (Ballantyne *et al.* 2011: 775). In a similar light, McKie (2007) argued that educational programming as well as the pro-poor or pro-conservation benefits of last chance tourism might outweigh the potential environmental impacts. Last chance tourism could also lead to more environmentally conscious holiday choices by travellers through carbon-labelling of travel packages (see also Gössling 2011 for additional information on mitigating the impacts of climate change from tourism).

Amongst the numerous potential benefits of last chance tourism there are also important ethical dilemmas that require attention. Lemieux and Eagles (this volume) for example ask how much damage to the natural environment should be incurred through economic needs that may be associated with last chance tourism. Lemelin and Baikie (this volume) suggest that ultimately the answer to these and other questions will come as a judgement that must be made by decision makers, local communities, and tourism stakeholders. Tourism operators and managers will increasingly have to make these decisions regarding the development of last chance tourism opportunities and will need to balance the environmental and socio-economic trade-offs. Coghlan (this volume) for example, criticizes the trade-off currently being made between perpetuating an inflated image of vulnerability in order to generate much-needed conservation funds for the Great Barrier Reef and promoting a healthy reef ecosystem that is more desirable for tourists.

Perpetuating an exaggerated image of vulnerability for the purpose of profit is one of the potential dangers of the last chance tourism concept and is something that requires careful ethical consideration and further critical investigation. The exaggerated perpetuation of tourism resource vulnerability is regularly in direct conflict with the desires and opinions of many tourism operators. For example, Lemelin and Baikie (this volume) found that local and regional tourism operators have decided to challenge the 'dying aspect' of their sites and cultures and rather choose to promote and market their 'wellness' and resilience. Steiger *et al.* (this volume) reiterate this perspective suggesting that operators of alpine ski areas located in vulnerable regions prefer to market an image of 'absolute snow' in contrast to the notion of last chances. Perhaps the most acute example of tourism operators protesting the notion of last chance tourism was from polar bear viewing operators in Churchill, Manitoba who in an online public forum discussing the notion clearly and swiftly rejected it stating: 'On behalf of our industry I submit that Churchill tourism operators don't participate in this discussion [of last chance tourism]' (Ord 2010).

The common rejection of last chance tourism by tourism operators is not particularly surprising considering their likely desire to remain economically sustainable now and into the future. Promoting the vulnerability of the platform or foundation of which operators have built their businesses admittedly seems contradictory. The disconnect that clearly exists between the extent to which last chance tourism is actually occurring on the ground and the disproportionate amount of attention the idea is getting in the media is concerning. However, as noted by several authors in this volume, there does seem to be a demand for last chance tourism experiences, and under increasing economic pressures associated with global climate and other social change it is highly plausible that more and more operators will begin, or in some cases continue, to capitalize on the existence of demand for last chance tourism experiences.

## Future directions and final remarks

In an attempt to build on the foundational research showcased in this volume, we felt it pertinent to identify key research areas that will need future attention in order to more fully understand the newly emerging phenomenon of last chance tourism. The most pressing research needs are summarized below:

- In order to better understand the scale, scope and nature of last chance tourism, geographically diverse case studies beyond those presented in this volume need to be conducted. Longitudinal research is critical to allow exploration of changes over time that could be rapid and turbulent.
- Last chance tourists, some would claim, can act as 'agents of change' or advocates or ambassadors. However, as Lamers *et al.* (this volume) hints, ambassadorship is poorly understood, and research is required to assess whether this notion is anything other than 'wishful thinking' (Bauer 2001).
- As the chapters by Olsen *et al.* and Hvenegaard illustrate, showcasing research on the world wide web has taken discussion, previously restricted to academic journals and specialist interest groups, into the public sphere. Yet various marketing approaches promoting last chance tourism, as Coghlan and Frew (both this volume) explain, have been and continue to be showcased. It is evident that a disconnect exists between the media, researchers, local operators and managers. Understanding more fully this disconnect may lead to better engagement between these groups.
- Lemieux and Eagles, and Steiger *et al.* (both this volume) explain that there is little understanding of how climate change may stimulate a diversification of last chance tourism products and destinations, on the one hand, and affect visitor demand and visitor expériences, on the other. They suggest that understanding the potential behavioural response of tourists to the increasing or fluctuating vulnerability of a tourism resource is an important first step towards developing and implementing adaptation strategies aimed at sustaining the resource being utilized while simultaneously taking advantage of economic opportunities. Similarly, the basic understanding of adaptation to last chance tourism is poorly documented, and as Stewart *et al.* (this volume) highlight, assessing adaptation responses to last chance tourism at multiple scales will be important. Furthermore the suggestion by Johnston *et al.* (this volume) that last chance tourism may in parallel present first chance tourism opportunities (i.e. as an adaptation to last chance tourism) deserves further elaboration.
- The appeal of vanishing landscapes and disappearing wildlife icons have been cornerstones of this volume (see Newsome and Rodger; Stewart *et al.*); however, as identified by Baikie and Lemelin (this volume) and noted by Johnston *et al.* (this volume), it is critical to recognize the socio-cultural dimensions of last chance tourism and, as such, this needs to be addressed in future research. The connections identified between last chance tourism (ecocide) and dark tourism (death appeal) also warrant additional reflection.

- Drawing on the contributions by Smith, and Shelton and Lemieux and Eagles (all this volume), the philosophical and ethical dimensions of last chance tourism requires further analysis. For example, managing what species or other valued resources could or should be saved and the obligations to 'future generations' who will have to live with, and possibly bear the costs associated with, management decisions made today clearly requires further deliberation and examination.
- The role of local stakeholders and communities in last chance tourism, while discussed in some chapters (see Johnston *et al.*, Lemelin and Baikie both this volume) demands further elaboration.

Despite their role in over-publicizing a trend which has only recently gained some traction, it would be erroneous to solely blame the media for creating the last chance tourism hype. Many stakeholders (including scientists, environmental groups, researchers, and certain tourism operators) have touted the alarming aspects of climate change and, in some cases knowingly and in others unknowingly, promoted the vulnerability of last chance species and sites. There is also an obvious increase in desire amongst some individuals to view vulnerable and 'dying' destinations, wildlife, and culture.

We can not deny the emergence of a last chance tourism phenomenon, but we must not fear it either. Considering the economic potential of last chance tourism for individuals and communities, which are often located in remote and less prosperous areas of the world, Kendle (2008) asks 'can we really blame anyone for jumping on the doomsday bandwagon?' The potential environmental and conservation benefits of last chance tourism are another reason to embrace its potential. Many conservationist and protectionist movements that have stemmed directly from the promotion and politicization of vulnerable flora and fauna have played prominent roles in reversing their dying fate. Twenty years after Douglas Adams and Mark Carwardines' radio broadcast 'last chance to see' was aired, Mark Carwardine along with Stephen Fry relived the original journey, which took them all over the world in search of endangered animals (a book based on the original radio show was published in 1990 – see Adams and Carwardine 1990). Adams and Carwardines' original travelogue highly publicized nine vulnerable wildlife species including the Komodo Dragon, Mountain Gorilla, Northern White Rhino, Manatee, Kakapo, Fur Seal, Aye-aye, Yangtze River Dolphin and Rodrigues fruit bat. Although the fate of these species has been mixed with the northern white rhino and the Yangtze river dolphin now 'functionally extinct' (Fry 2009: 9) to the contrary, many of the species originally documented have been downgraded on the IUCN conservation status classification.

This example of showcasing and preserving vulnerable species reminds us of the importance of generating a call to action and the potential positive power of witnessing and experiencing vulnerable species and landscapes. As outlined by Fry (2009: 10) 'Let us never stop talking about the creatures we share the planet with. The first step is to know them a little better.' As was the case with Douglas Adams' list of vulnerable species, it is our hope that 20 years from now the promotion and visitation of the destinations, wildlife species, and cultures profiled in this book will assist in their conservation. With any luck the media headlines of the future will read, 'it is all still here: last chance tourism was an exaggeration'.

## References

Aall, C. and Høyer, K.G. (2005) 'Tourism and climate change adaptation: the Norwegian case', in C.M. Hall and J. Higham (eds.) *Tourism, recreation and climate change*, Clevedon, UK, and Buffalo, NY: Channel View Publications.

Adams, D. and Carwardine, M. (1990) *Last Chance to See*, London: Heinemann.

Ballantyne, R., Packer, J., and Sutherland, L.A. (2011) 'Visitors' memories of wildlife tourism: Implications for the design of powerful interpretive experiences', *Tourism Management* 32: 770–779.

Bauer, T.G. (2001) *Tourism in the Antarctic: Opportunities, constraints and future prospects*, NewYork: Haworth Hospitality Press.

Dawson, J., Stewart, E.J., Lemelin, H., and Scott, D. (2010) 'The carbon cost of polar bear viewing in Churchill, Canada', *Journal of Sustainable Tourism* 18: 319–336.

Dawson, J., Johnston, M.E., Stewart, E.J., Lemieux, C.J., Lemelin, H., and Maher, P. (in press) 'Ethical considerations of last chance tourism', *Journal of Ecotourism.*

Eijgelaar, E., Thaper, C., and Peeters, P. (2010) 'Antarctic cruise tourism: the paradoxes of ambassadorship, "last chance tourism" and greenhouse gas emissions', *Journal of Sustainable Tourism* 18: 337–354.

Frew, E.A. (2008) 'Climate change and doom tourism: Advertising destinations 'before they disappear', in J. Fountain and K. Moore (eds.), *Recreating Tourism: Proceedings of the New Zealand Tourism & Hospitality Research Conference 2008*. Canterbury, New Zealand: CD Rom, Lincoln University.

Fry, S. (2009) 'Foreword', in Carwardine, M., *Last Chance to See: In the footsteps of Douglas Adams*, London: Collins.

Gössling, S. (2011) *Carbon Management in Tourism: Mitigating the Impacts on Climate Change*, New York: Routledge.

Hamilton, J.M., Maddison, D.J., and Tol, R.S.J. (2005) 'Climate change and international tourism: a simulation study', *Global Environmental Change* 15: 253–266.

Intelligent Travel (2007, December) *Climate change voyeurism?* Online: http://blogs.nationalgeographic.com/blogs/intelligenttravel/2007/12/climate-change-voyeurism.html (accessed 21 December 2007).

Iso-Ahola, S.E. (1986) 'A theory of substitutability of leisure behavior', *Leisure Sciences* 8(4): 367–389.

JWT (2007) *80 Things to watch in 2008.* Online: www.influxinsights.com/blog/article/1724/jwt-has-80-things-for-us-to-watch-in-08.html (accessed 21 December 2007).

Kendle, A. (2008) 'Doomsday tourism: seeing it before you can't', *Vagabondish: the travelzine for today's vagabond.* Online: www.vagabondish.com/environment-doomsday-tourism-travel-tours/ (accessed 1 October 2008).

Kirby, A. (2002) 'Hope rise for mountain gorillas'. *BBC News World Edition.* Online: news.bbc.co.uk/2/hi/science/nature/2332527.stm (accessed 5 May 2011).

Leahy, S. (2008) 'Climate change: tourism at the end of the world', *IPS*, Online: http://ipsnews.net/news.asp?idnews=40837 (accessed 1 October 2008).

Lemelin, R.H. (2006) 'The gawk, the glance, and the gaze: Occular consumption and polar bear tourism in Churchill, Manitoba Canada', *Current Issues in Tourism* 9(6): 516–534.

McCarthy, M. (2007) 'A holiday at the end of the earth: tourists paying to see global warming in action', *The Independent*. Online: www.independent.co.uk/environment/climate-change/a-holiday-at-the-end-of-the-earth-tourists-paying-to-see-global-warming-in-action-447206.html (accessed 1 October 2008).

McKie, R. (2007) '10 wonders of the vanishing world', *The Observer*. Online: www.guardian.co.uk/travel/2007/jan/28/wildlifeholidays.ethicalholidays.green (accessed 1 October 2008).

Ord, C. (2010) 'Last chance tourism? Vanishing destinations and tourism's role in saving them'. Online: www.yourtravelchoice.org/2010/12/last-chance-tourism-vanishing-destinations-and-tourisms-roles-in-saving-them/ (accessed 2 April 2011).

Ruiz, R. (2008) 'Endangered natural wonders worth seeing', *Forbs*. Online: www.forbes.com/2008/09/15/travel-endangered-ecosystems-forbeslife-cx_rr_0915travel.html (accessed 1 October 2008).

Salkin, A. (2007) 'Before it disappears', *The New York Times*. Online: www.nytimes.com/2007/12/16/fashion/16dissapear.html?ei=5124&en=aaa1743bd1 (accessed 1 October 2008).

Scott, D., Dawson, J., and Jones, B. (2008) 'Climate change vulnerability of the US Northeast winter recreation-tourism sector', *Mitigation and Adaptation Strategies for Global Change* 13: 577–596.

Shipman, T. (2007) 'Nature's 'doom' is tourist boom', *Daily Telegraph*. Online: www.asoc.org/Portals/0/Natures%20doom%20is%20tourist%20boom%20122307.pdf (accessed 1 October 2008).

Stewart, E.J., Dawson, J.D., and Draper, D. (2011) 'Monitoring patterns of cruise tourism across Arctic Canada', in M. Luck, P.T Maher and E.J. Stewart (eds.) *Cruise Tourism in the Polar Regions: Promoting Environmental and Social Sustainability*. London: Earthscan.

Todras-Whitehill, E. (2007) 'As ice recedes interest surges', *The New York Times*. Online: http://travel.nytimes.com/2007/12/09/travel/09adventure.html (accessed 1 October 2008).

# Index